SIGNIFICANT CHANGES TO THE

INTERNATIONAL PLUMBING CODE®
INTERNATIONAL MECHANICAL CODE®
INTERNATIONAL FUEL GAS CODE®

2021 EDITION

SIGNIFICANT CHANGES TO THE INTERNATIONAL PLUMBING CODE® INTERNATIONAL MECHANICAL CODE® INTERNATIONAL FUEL GAS CODE®

2021 EDITION

International Code Council

ICC Staff:

Executive Vice President and Director of Business Development:
Mark A. Johnson

Senior Vice President, Business and Product Development:
Hamid Naderi

Vice President and Technical Director, Products and Services:
Doug Thornburg

Senior Marketing Specialist:
Dianna Hallmark

ISBN: 978-1-952468-21-6

Project Manager: Steve Van Note

Publications Manager: Anne F. Kerr

Production Technician: Dianna Logan

Cover Design: Ricky Razo

INTERNATIONAL CODE COUNCIL®

Errata on various ICC publications may be available at www.iccsafe.org/errata.

First Printing: October 2020

PRINTED IN THE USA

T026928

Contents

SIGNIFICANT CHANGES TO THE International Plumbing Code

SIGNIFICANT CHANGES TO THE International Mechanical Code

SIGNIFICANT CHANGES TO THE International Fuel Gas Code

Preface

The purpose of *Significant Changes to the International Plumbing Code®, International Mechanical Code®, International Fuel Gas Code®* 2021 Edition is to familiarize plumbing and mechanical officials, building officials, fire officials, plans examiners, inspectors, design professionals, contractors and others in the building construction industry with many of the important changes to these codes. This publication is designed to assist code users in identifying the specific code changes that have occurred and, more importantly, in understanding the reasons behind the changes. It is also a valuable resource for jurisdictions in their code-adoption process.

Only portions of the total number of code changes to the *International Plumbing Code, International Mechanical Code* and *International Fuel Gas Code* are discussed in this book. The changes selected were identified for a number of reasons, including their frequency of application, special significance or change in application. However, the importance of the changes not included is not to be diminished. Further information on all code changes can be found in the *Complete Revision History to the 2021 I-Codes: Successful and Public Comments,* available in 2021 from the International Code Council® (ICC®) online store, http://shop.iccsafe.org. This resource collection provides the published documentation for each successful code change contained in the 2021 *International Plumbing Code* and *International Mechanical Code*.

Throughout this *Significant Changes* book, each change is accompanied by a photograph or an illustration to assist in and enhance the reader's understanding of the specific change. A summary and a discussion of the significance of the change are also provided. Each code change is identified by type, be it an addition, modification, clarification or deletion.

The code change itself is presented in a legislative format similar to the style utilized for code-change proposals. Deleted code language is shown with a strikethrough, and new code text is indicated by underlining. As a result, the actual 2021 code language is provided as well as a comparison with the 2018 language, so the user can easily determine changes to the specific code text.

As with any code-change text, *Significant Changes to the International Plumbing Code, International Mechanical Code, International Fuel Gas Code* 2021 Edition is best used as a companion to the 2021 IPC, 2021 IMC and 2021 IFGC. Because only a limited discussion of each change is provided, the reader should reference the code itself to gain a more comprehensive understanding of the code change and its application.

The commentary and opinions set forth in this text are those of the authors and do not necessarily represent the official position of ICC. In addition, they may not represent the views of any enforcing agency, as such agencies have the sole authority to render interpretations of the IPC, IMC and IFGC. In many cases, the explanatory material is derived from the reasoning expressed by code-change proponents.

Comments concerning this publication are encouraged and may be directed to ICC at *significantchanges@iccsafe.org*.

About the *International Plumbing Code, International Mechanical Code* and *International Fuel Gas Code*

Code officials, design professionals, contractors and others involved in the building construction industry recognize the need for modern, up-to-date codes addressing the design and installation of plumbing, mechanical and fuel gas systems through both prescriptive and performance requirements. The 2021 editions of the *International Plumbing Code* (IPC®), *International Mechanical Code* (IMC®) and *International Fuel Gas Code* (IFGC®) are intended to meet these needs through model code regulations that safeguard the public health and safety in all communities, large and small. The IPC, IMC and IFGC are kept up to date through ICC's open code-development process. The provisions of the 2018 editions, along with those code changes approved through 2019, make up the 2021 editions.

The IPC, IMC and IFGC are three codes in a family of International Codes® published by ICC. These comprehensive codes establish minimum regulations for plumbing, mechanical and fuel gas systems by means of prescriptive and performance-related provisions, and are founded on broad-based principles that make possible the use of new materials and new system designs. The IPC, IMC and IFGC are available for adoption and use by jurisdictions internationally. Their use within a governmental jurisdiction is intended to be accomplished through adoption by reference, in accordance with proceedings establishing the jurisdiction's laws.

Acknowledgments

Fred Grable, author of the IPC section, thanks all of the people who have called him over the years with questions about the requirements in the many editions of the IPC. The conversations have led to his greater understanding of how the codes are applied in the built environment and have enabled him to better help others.

Gregg Gress, author of the IMC and IFGC portions of this book, thanks fellow ICC staff, the graphic artists and all those who donated photographs.

About the Authors

Fred Grable, P.E.

International Code Council

Senior Staff Engineer, Technical Services

Fred Grable, ICC Senior Staff Engineer and former master plumber, brings extensive experience as both a plumbing contractor and engineer. He has nearly two decades of experience as an engineer for an industrial pressure piping manufacturer and in wastewater treatment equipment field services. He also founded his own plumbing company and managed new construction, remodeling, and repair of residential and commercial plumbing and gas systems. Fred is a registered professional engineer in two states. He has worked for the ICC for over ten years in the Codes and Standards Department and is the Secretariat for the *International Plumbing Code* (IPC) and the plumbing portion of the *International Residential Code®* (IRC®), the *International Private Sewage Disposal Code®* (IPSDC®), the *International Swimming Pool and Spa Code®* (ISPSC®) and the residential provisions of the *International Energy Conservation Code®* (IECC®).

Gregg Gress

International Code Council

Senior Technical Staff, Technical Services

Gregg Gress is a former Senior Technical Staff member of the ICC and was responsible for the maintenance of the *International Mechanical Code* (IMC) and *International Fuel Gas Code* (IFGC), the commercial portion of the *International Energy Conservation Code* (IECC) and the mechanical and electrical parts of the *International Residential Code* (IRC). He is also an author of the IMC, IRC and IFGC Commentary books and has provided training and code interpretations of the ICC codes. Gregg was formerly a code official for a municipality and has worked in the plumbing, mechanical/HVAC and electrical trades and has been employed by BOCA and the ICC for 25 years. Gregg has served on multiple AGA, NFPA and ASHRAE project committees.

About the International Code Council®

The International Code Council is a nonprofit association that provides a wide range of building safety solutions including product evaluation, accreditation, certification, codification and training. It develops model codes and standards used worldwide to construct safe, sustainable, affordable and resilient structures. ICC Evaluation Service (ICC-ES) is the industry leader in performing technical evaluations for code compliance, fostering safe and sustainable design and construction.

Washington DC Headquarters:
500 New Jersey Avenue, NW, 6th Floor, Washington, DC 20001

Regional Offices:
Eastern Regional Office (BIR)
Central Regional Office (CH)
Western Regional Office (LA)

Distribution Center (Lenexa, KS)

888-ICC-SAFE (888-422-7233)
www.iccsafe.org

Family of Solutions:

ICC INTERNATIONAL CODE COUNCIL® · ES ICC EVALUATION SERVICE® · IAS INTERNATIONAL ACCREDITATION SERVICE® · GENERAL CODE · ICC SKGA S. K. GHOSH ASSOCIATES · Alliance for National & Community Resilience · NTA® · ICC COMMUNITY DEVELOPMENT SOLUTIONS

SIGNIFICANT CHANGES TO THE

INTERNATIONAL PLUMBING CODE®

2021 EDITION

PART 1

International Plumbing Code

Chapters 1 through 15

- **Chapter 1** Scope and Administration
  No changes addressed
- **Chapter 2** Definitions
- **Chapter 3** General Regulations
- **Chapter 4** Fixtures, Faucets and Fixture Fittings
- **Chapter 5** Water Heaters
- **Chapter 6** Water Supply and Distribution
- **Chapter 7** Sanitary Drainage
- **Chapter 8** Indirect/Special Waste
  No changes addressed
- **Chapter 9** Vents
- **Chapter 10** Traps, Interceptors and Separators
- **Chapter 11** Storm Drainage
- **Chapter 12** Special Piping and Storage Systems
- **Chapter 13** Nonpotable Water Systems
- **Chapter 14** Subsurface Landscape Irrigation Systems No changes addressed
- **Chapter 15** Referenced Standards
  No changes addressed

Chapter 1 of the *International Plumbing Code*® (IPC®) clarifies how the code will be enforced by code officials. Chapter 2 contains definitions of plumbing code terminology. General regulations in Chapter 3 identify requirements not listed in other code chapters, such as testing and inspections. Fixtures and water heaters are addressed in Chapters 4 and 5, respectively. Chapters 6 and 7 regulate water and drainage piping systems. Indirect and special waste is covered in Chapter 8. Chapter 9 details acceptable venting methodologies with in-depth provisions for piping arrangements. Chapter 10 contains the provisions for traps and various receptors. Storm drainage, with its collection system piping provisions, is covered in Chapter 11. Installation, design, storage, handling and use of nonflammable medical gas systems are addressed in Chapter 12. Nonpotable water systems such as those for the storage, treatment, and use of gray water, rainwater, reclaimed water and alternate on-site nonpotable water are addressed in Chapter 13. Methods for the use of nonpotable water for subsurface irrigation are addressed in Chapter 14. Standards referenced by the code sections are indicated in Chapter 15 along with specific details about the applicable edition year and title. Appendices A through E cover nonmandatory provisions for permit fees, rainfall rates, degree design temperature, water (piping) sizing methods and structural integrity protection rules for the notching of and boring of holes in wood and steel members. New Appendix F establishes rules for a board of appeals. ■

202

Definition of Copper Alloy

202

Definition of Public and Private

202

Definition of Water Dispenser

308.9

Bundled Hot Water Piping Insulation

403.1.1

Fixture Quantity Calculations for Multiple User Facilities

403.1.2

Fixtures in Single-user Toilet Rooms Count Towards Total Required Quantities

403.2

Multiple-user Nonseparated Toilet Facilities

403.3.1, 403.5

Facilities and Drinking Fountains on Accessible Routes

403.3.3

Group S Toilet Facility Location

403.6

Service Sink Location

405.4.3

Wall Hung Fixture Carrier Standard for Water Closets

407.2

Bathtubs are Not Required to have Overflow Outlets

410.3.2

Quantities of Standing versus Wheelchair Drinking Fountains

410.4

Drinking Fountain Substitution using Water Dispensers

411.3

Water Heaters for Emergency Showers and Eye Wash Stations

412.3

Shower Control Valves to be Rated for the Installed Shower Head

412.5

Methods for Limiting Water Temperature Discharged to Bathtubs

412.10

Methods for Temperature Limitation at Head Shampoo Sinks and Footbaths

419.5

Tempered Water for Public Hand-Washing Lavatories

421.3.1

Standard for Shower Waste Fittings

501.2

ASSE 1017 Temperature Actuated Mixing Valves

602.3.5

Potable Water Pumps to Comply with NSF 61

605.12.3, 605.13.6

Solder and Flux to Conform to NSF 61

606.1

Individual Tenant Water Shut-off Valve

607.1.1, 607.1.2

Water Heaters Providing Tempered Water to Fixtures

608.15.2.1

Discharge from Backflow Preventer Relief Opening

608.17.2

Backflow Device for Low Hazard Boiler Applications

609.2

Two Water Service Pipes for Group I-2, Condition 2 Healthcare Facilities

609.2.1

Tracer Wire for Buried Nonmetallic Water Service Piping

TABLE 702.3

ABS Building Sewer Pipe Standard

705.2.4, 705.10.4

PVC and ABS Push-fit DWV Fittings

708.1.6

Removable Fixture Traps Serving as Cleanouts

717, 718

Methods for Restoring Building Sewer Piping

903.1.3

Protected Outdoor Vent Termination Method

915.1

Food Waste Disposers on Combination Waste and Vent Systems

1002.1

Direct Connection to Hydromechanical Grease Interceptor

1002.4.1.5

Fixture Drains Serving as a Trap Priming Method

1102.6

Roof Drains to be Tested and Rated for Flow

1106.2.1

Rainfall Rate Conversion Method

1202.1

Nonflammable Medical Gas Systems

1301.1

Nonpotable Rainwater Standard Alternative

202

Definition of Copper Alloy

CHANGE TYPE: Clarification

CHANGE SUMMARY: The term "copper alloy" is used in numerous locations in the code text. The new definition broadly describes the entire range of copper alloy materials that a manufacturer might use for a product.

2021 CODE: **COPPER ALLOY.** A metal alloy where the principle component is copper.

CHANGE SIGNIFICANCE: The addition of this definition to the code is a simple clarification to support the use of the term "copper alloy" throughout the code. Long ago, the term for alloys of copper used in the plumbing industry was "brass" (e.g., typically red brass or yellow brass). However, brasses are only one group of a wider array of copper alloys that a manufacturer might wish to use for products. In 1974, a unified numbering system (UNS) was created to classify copper alloys. Because the term "brass" may not necessarily be technically correct for all products, "copper alloy" is used to cover all possible materials that are alloys of copper.

Unified Numbering System Ranges for Copper Alloys

Wrought Products

C100xx-C150xx Cu Commercially Pure Copper

C151xx-C199xx Age Hardenable Cu (w/ Cd, Be, Cr, Fe)

C2xxxx Cu-Zn alloys Brasses

C3xxxx Cu-Zn-Pb alloys Leaded brasses

C4xxxx Cu-Zn-Sn alloys Tin bronzes

C5xxxx Cu-Sn and Cu-Sn-Pb Phosphor bronze alloys

C6xxxx Cu-Al and Cu-Si Bronzes

C7xxxx Cu-Ni Copper Nickel and Cu-Ni-Zn Nickel Silver

Cast Products

C800xx-C811xx Commercially Pure Coppers

C813xx-C828xx 95-99% Copper

C833xx-C899xx Cu-Zn alloys containing Sn, Pb, Mn, or Si

C9xxxx Other alloys, including tin bronze, aluminum bronze, copper nickel

202

Definition of Public and Private

CHANGE TYPE: Modification

CHANGE SUMMARY: The definitions of "public" and "private" are simplified to make a clearer distinction as to which plumbing fixtures are intended to be configured for public use.

2021 CODE: PUBLIC OR PUBLIC UTILIZATION. In the classification of plumbing fixtures, "public" applies to fixtures ~~in general toilet rooms of schools, gymnasiums, hotels, airports, bus and railroad stations, public buildings, bars, public comfort stations, office buildings, stadiums, stores, restaurants and other installations where a number of fixtures are installed so that their utilization is similarly unrestricted~~ with unrestricted exposure to walk-in traffic.

PRIVATE. In the classification of plumbing fixtures, "private" applies to fixtures ~~in residences and apartments, and to fixtures in nonpublic toilet rooms of hotels and motels and similar installations in buildings where the plumbing fixtures are intended for utilization by a family or an individual~~ that are not public.

CHANGE SIGNIFICANCE: The definition of "public" is aligned with that of standards ASME A112.18.1, *Plumbing Supply Fittings*, and ASHRAE 90.1, *Energy Standard for Buildings*, to minimize the number of lavatory fixture fittings (faucets) that must comply with the code's public use requirements for 1) discharging only tempered water and 2) having a flow rate of not greater than 0.5 gpm.

Public lavatory faucets are the only plumbing fixtures required by the code to discharge tempered (between 85 and 110°F) water. The revised definitions have little, if any, impact on public lavatories. However, as the terms "public" and "private" are used throughout the code for purposes beyond identifying lavatory types, the revised definitions could give new meaning to other code sections even though the approved proposal may not have intended to change the meaning in those contexts.

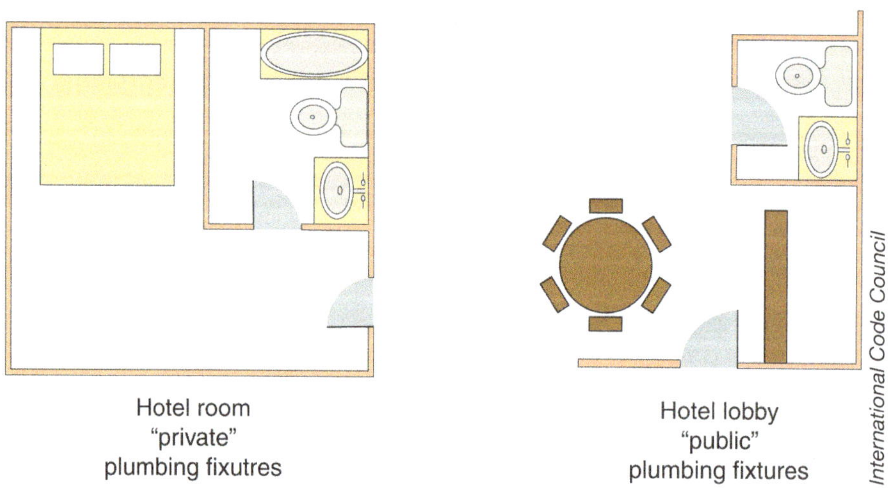

Public versus private uses

CHANGE TYPE: Modification

CHANGE SUMMARY: A bottled water unit is no longer defined by the code as a water dispenser.

2021 CODE: **WATER DISPENSER.** A plumbing fixture that is manually controlled by the user for the purpose of dispensing potable drinking water into a receptacle such as a cup, glass or bottle. Such fixture is connected to the potable water distribution system of the premises. ~~This definition includes a freestanding apparatus for the same purpose that is not connected to the potable water distribution system and that is supplied with potable water from a container, bottle or reservoir.~~

CHANGE SIGNIFICANCE: Section 410.4 uses the term "water dispenser" concerning a substitution for drinking fountains. The definition change results in the elimination of bottled water units and other self-contained potable water dispensing apparatus as a code-sanctioned alternative to some number of drinking fountains. Although the proposal's reason statement did not mention this impact, referencing the term's revised definition when interpreting Section 410.4 may lead the reader to that conclusion.

202

Definition of Water Dispenser

International Code Council

Bottled water unit is not a water dispenser

308.9

Bundled Hot Water Piping Insulation

CHANGE TYPE: Modification

CHANGE SUMMARY: Where hot water piping for a manifold system is in a bundle with cold water piping, the hot water piping is required to be insulated but not necessarily individually insulated.

2021 CODE: 308.9 Parallel water distribution systems. Piping bundles for manifold systems shall be supported in accordance with Table 308.5. Support at changes in direction shall be in accordance with the manufacturer's instructions. Where hot water piping is bundled with cold ~~or hot~~ water piping~~, each~~ hot water ~~pipe~~ piping shall be insulated in accordance with Section 607.5.

CHANGE SIGNIFICANCE: The bundling of piping from manifold systems is often required very near the manifold. Individually insulating each hot water pipe (tube) in these bundles results in very large openings needed in framing members for the bundles to pass through. In those situations, the bundle of hot water piping can be insulated. The reference to Section 607.5 directs the code user to the requirements for pipe insulation.

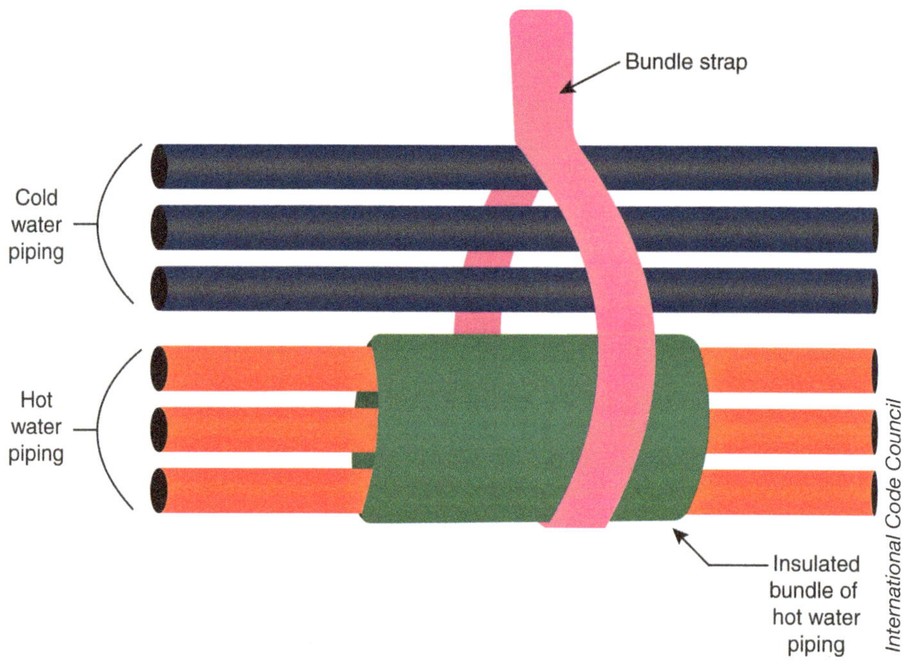

Hot water piping bundled with cold water piping

CHANGE TYPE: Modification

CHANGE SUMMARY: The minimum fixture quantities for multiple-user toilet facilities designed to serve all genders must be calculated 100 percent based on total occupant load.

403.1.1

Fixture Quantity Calculations for Multiple User Facilities

2021 CODE: **403.1.1 Fixture calculations.** To determine the occupant load of each sex, the total occupant load shall be divided in half. To determine the required number of fixtures, the fixture ratio or ratios for each fixture type shall be applied to the occupant load of each sex in accordance with Table 403.1. Fractional numbers resulting from applying the fixture ratios of Table 403.1 shall be rounded up to the next whole number. For calculations involving multiple occupancies, such fractional numbers for each occupancy shall first be summed and then rounded up to the next whole number.

Exceptions:

1. The total occupant load shall not be required to be divided in half where approved statistical data indicate a distribution of the sexes of other than 50 percent of each sex.
2. Where multiple-user facilities are designed to serve all genders, the minimum fixture count shall be calculated 100 percent, based on total occupant load. In such multiple-user facilities, each fixture type shall be in accordance with ICC A117.1 and each urinal that is provided shall be located in a stall.
3. Distribution of the sexes is not required where single-user water closets and bathing room fixtures are provided in accordance with Section 403.1.2.

Plumbing Fixture Calculations

Multiple-User Facility for Serving All Genders

Given: Business Use having an Occupant Load of 60. Toilet facility design chosen to be one multiple-user facility to serve all genders.

Per Table 403.1:
Water Closet Ratio: 1 per 25 for the first 50 and 1 per 50 for the remainder exceeding 50
Lavatory Ratio: 1 per 40 for the first 80 and 1 per 80 for the remainder exceeding 80

Calculations:
WCs: 50/25 + (60 − 50)/50 = 2.2 Round up to 3.
LAVs: 60/40 = 1.5 Round up to 2.

Single-User Facilities

Given: Business Use having an Occupant Load of 60. Toilet facility design chosen to be all single-user facilities.

Per Table 403.1:
Water Closet Ratio: 1 per 25 for the first 50 and 1 per 50 for the remainder exceeding 50
Lavatory Ratio: 1 per 40 for the first 80 and 1 per 80 for the remainder exceeding 80

Calculations:
WCs: 50/25 + (60 − 50)/50 = 2.2 Round up to 3.
LAVs: 60/40 = 1.5 Round up to 2. However, because each single-user facility requires a LAV, the required number of LAVs = 3.

CHANGE SIGNIFICANCE: The new Exception 2 introduces the concept of multiple-user toilet facilities designed to serve all genders. In the context of this exception, gender does not mean sex. Such multiple-user facilities are available for any person to use, regardless of their gender identity. Along with Exception 6 of Section 403.2, additional requirements for this type of facility are few. Many arrangements are possible and other than urinal locations, additional privacy features, beyond what is already required in any multiple-user toilet room intended for use by the same sex, are not required. The design need only serve all genders.

The requirement, "the minimum fixture count shall be calculated 100 percent, based on the total occupant load," attempts to override the requirement of the base section for dividing the total occupant load of the building in half (assuming a 50-50 male/female distribution) when applying the fixture ratios of Table 403.1. Calculations are straight forward where the male and female fixture ratios in the table are identical. Where different table ratios are indicated, calculations will require adjustments to accommodate the difference.

Where a building is designed to have single-user toilet facilities (the subject of Section 403.1.2), and the male and female fixture ratios in the table are identical, Exception 2 (proposal P15) allows the calculations to be straight forward (i.e., distribution of sexes is not required). However, where different table ratios are indicated, calculations will require adjustments to accommodate the difference.

403.1.2

Fixtures in Single-user Toilet Rooms Count Towards Total Required Quantities

CHANGE TYPE: Clarification

CHANGE SUMMARY: The numbers of plumbing fixtures in all single-user toilet rooms and bathing rooms count towards the total number required for a building.

2021 CODE: **403.1.2 Single-user toilet ~~facility~~ and bathing room fixtures.** The plumbing fixtures located in single-user toilet ~~facilities~~ and bathing rooms, including family or assisted-use toilet and bathing rooms that are required by Section 1109.2.1 of the *International Building Code*, shall contribute toward the total number of required plumbing fixtures for a building or tenant space. Single-user toilet ~~facilities~~ and bathing rooms, and family or assisted-use toilet rooms and bathing rooms shall be identified as being available for use ~~either~~ by all persons regardless of their sex.

The total number of fixtures shall be permitted to be based on the required number of separate facilities or based on the aggregate of any combination of single-user or separate facilities.

CHANGE SIGNIFICANCE: The required number of plumbing fixtures for a building or space can be located in single-user facilities, separate facilities or a combination thereof. Given that single-user facilities serve all persons (not sex separated), the number of plumbing fixtures in those rooms count towards the total number of fixtures required for a building.

International Code Council

Fixtures in all single-user toilet rooms count toward total required

Fixtures in all single-user toilet rooms count toward total required

<u>**A mixture of Single-User toilet facilities and Multiple-User toilet facilities**</u>

Given: Business Use having an Occupant Load of 360. Design chosen to have 5 single-user toilet facilities and two multiple-user "separate sex" toilet facilities.

Per Table 403.1:
Water Closet Ratio: 1 per 25 for the first 50 and 1 per 50 for the remainder exceeding 50
Lavatory Ratio: 1 per 40 for the first 80 and 1 per 80 for the remainder exceeding 80

Calculations:
Occupant load for males = 360/2 = 180
Occupant load for females = 360/2 = 180

Males:
WCs: 50/25 + (180 − 50)/50 = 4.6 Round up to 5.
LAVs: 80/40 + (180 − 80)/40 = 4.5 Round up to 5.

Females:
WCs: 50/25 + (180 − 50)/50 = 4.6 Round up to 5.
LAVs: 80/40 + (180 − 80)/40 = 4.5 Round up to 5.

Total required WCs = 10
Total required LAVs = 10

Five single-user toilet facilities requires 5 WC and 5 Lavs. Five WCs and 5 Lavs remain for the multiple-user "separate sex" toilet facilities.

Distribute 2 WCs and 2 Lavs for male toilet facility and, 3 WCs and 3 LAVs for female toilet facility.

OR

Distribute 2 WCs and 2 Lavs for female toilet facility and, 3 WCs and 3 LAVs for male toilet facility.

403.2

Multiple-user Nonseparated Toilet Facilities

CHANGE TYPE: Modification

CHANGE SUMMARY: Designs for multiple-user facilities serving both sexes are possible.

2021 CODE: **403.2 Separate facilities.** Where plumbing fixtures are required, separate facilities shall be provided for each sex.

Exceptions:

1. through 4. remain unchanged.
5. Separate facilities shall not be required to be designated by sex where single-user toilets rooms are provided in accordance with Section 403.1.2.
6. Separate facilities shall not be required where rooms having both water closets and lavatory fixtures are designed for use by both sexes and privacy for water closets is provided in accordance with Section 405.3.4. Urinals shall be located in an area visually separated from the remainder of the facility or each urinal that is provided shall be located in a stall.

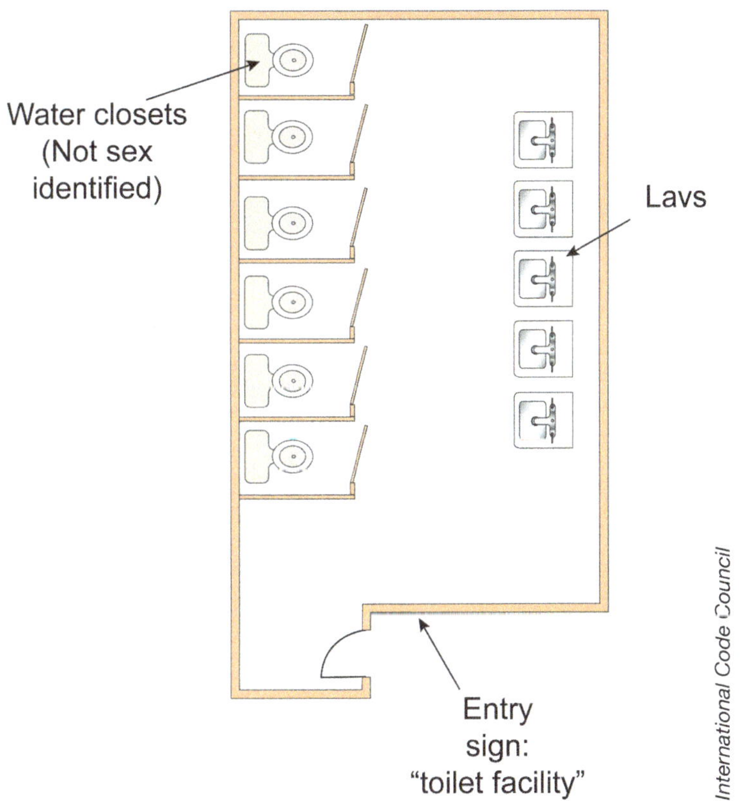

Multiple-user nonseparated toilet facilities

CHANGE SIGNIFICANCE: Exception 5 provides an allowance for single-user toilet rooms to not be designated by sex. However, for 2021, Section 403.1.2 requires that single-user toilet facilities not be designated by sex, making this exception unnecessary.

Exception 6 introduces the new concept of multiple-user toilet facilities designed to serve both sexes. Similar to the new concept introduced in Section 403.1.1 (multiple-user toilet facilities to serve all genders), this exception allows multiple-user facilities to be labeled for use by both sexes (male/female). As Section 405.3.4 was not changed for the 2021 edition, multiple-user separate sex toilet facility designs that were compliant with the 2018 code edition could comply with this exception, provided that any urinals are in stalls.

403.3.1, 403.5

Facilities and Drinking Fountains on Accessible Routes

CHANGE TYPE: Clarification

CHANGE SUMMARY: Because accessibility is covered by the requirements of Section 404 and those requirements include specifics concerning accessible routes, there is no need to repeat the accessible route requirement in these sections.

2021 CODE: **403.3.1 Access.** The route to the public toilet facilities required by Section 403.3 shall not pass through kitchens, storage rooms or closets. Access to the required facilities shall be from within the building or from the exterior of the building. ~~Routes shall comply with the accessibility requirements of the *International Building Code*.~~ The public shall have access to the required toilet facilities at all times that the building is occupied.

403.5 Drinking fountain location. Drinking fountains shall not be required to be located in individual tenant spaces provided that public drinking fountains are located within a distance of travel of 500 feet (152 m) of the most remote location in the tenant space and not more than one story above or below the tenant space. Where the tenant space is in a covered or open mall, such distance shall not exceed 300 feet (91 m). ~~Drinking fountains shall be located on an accessible route.~~

CHANGE SIGNIFICANCE: This change does not remove the requirement for toilet facilities and drinking fountains to be on accessible routes. Section 404 already properly covers this matter by referring to the *International Building Code* where Chapter 11 has full details (along with exceptions) about where accessible routes are and are not needed.

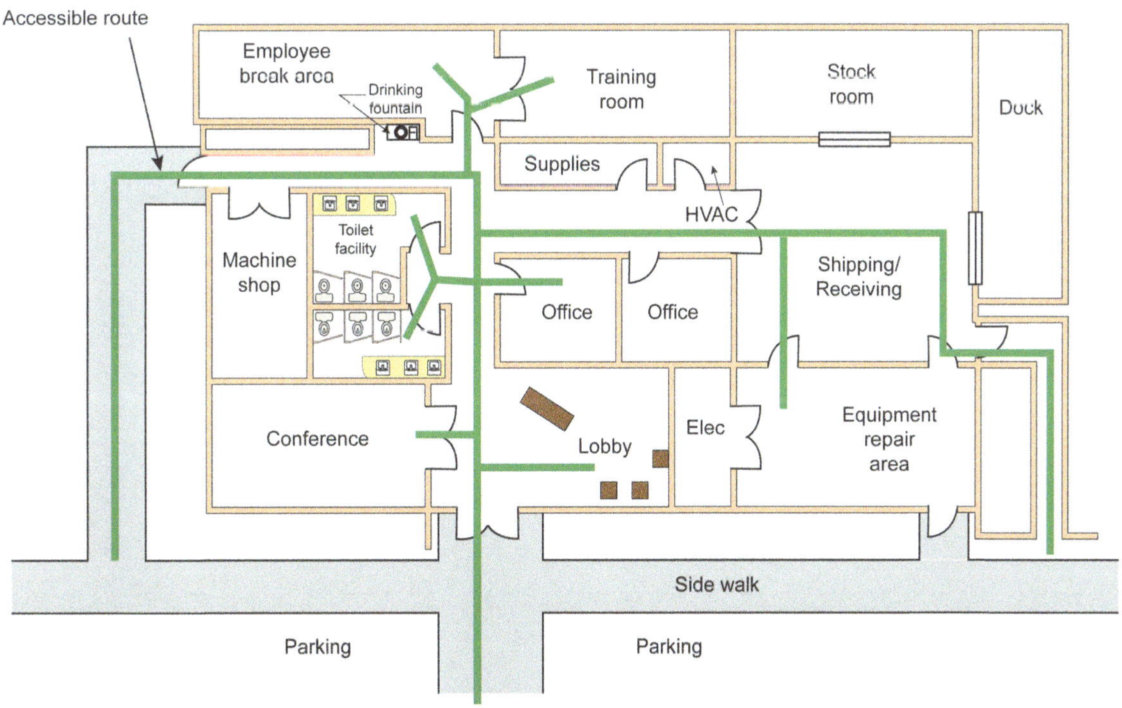

International Code Council

Accessible routes

403.3.3

Group S Toilet Facility Location

CHANGE TYPE: Modification

CHANGE SUMMARY: The location of toilet facilities in Group S occupancies can exceed the location and maximum distance limitations provided that the arrangement is approved.

2021 CODE: **403.3.3 Location of toilet facilities in occupancies other than malls.** In occupancies other than covered and open mall buildings, the required public and employee toilet facilities shall be located not more than one story above or below the space required to be provided with toilet facilities, and the path of travel to such facilities shall not exceed a distance of 500 feet (152 m).

Exceptions:

1. The location and maximum distances of travel to required employee facilities in factory and industrial occupancies shall be permitted to exceed that required by this section, provided that the location and maximum distances of travel are approved.
2. The location and maximum distances of travel to required public and employee facilities in Group S occupancies shall be permitted to exceed that required by this section, provided that the location and maximum distances of travel are approved.

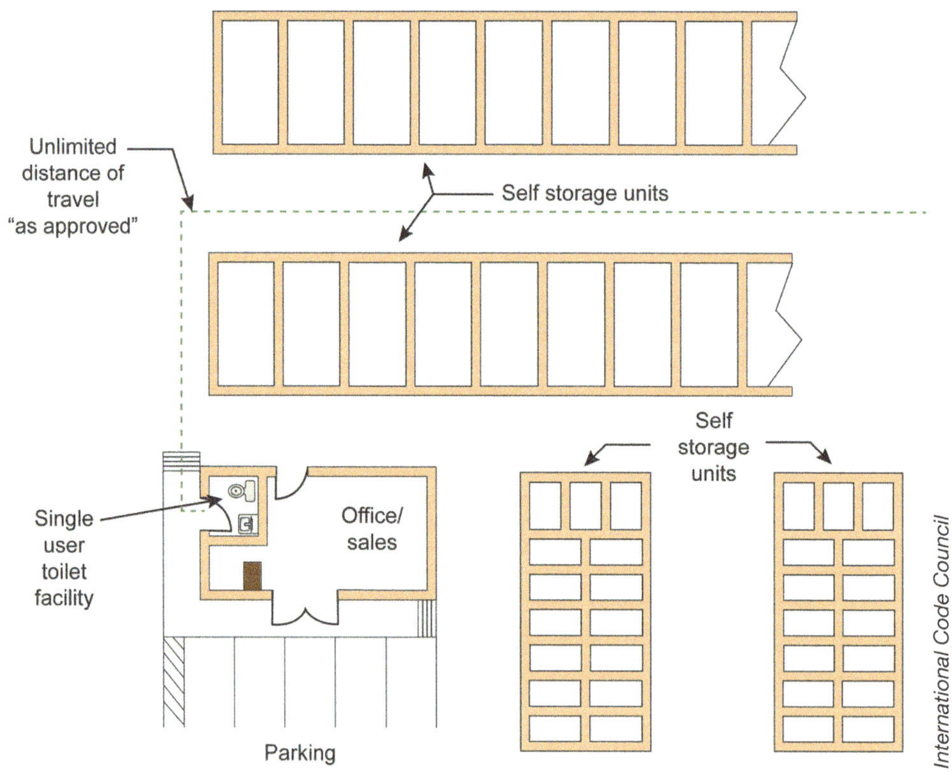

Storage facility with toilet facility at office

CHANGE SIGNIFICANCE: As some Group S occupancies have few people in them, it is not a cost-effective use of space or resources to require a duplication of toilet facilities simply because the maximum distance or number of stories was exceeded. One example is a large, multiple-level self-storage facility with an office at the facility entrance. Allowing the only employee (and public) toilet facility to be at the office is a common practice.

403.6

Service Sink Location

CHANGE TYPE: Addition

CHANGE SUMMARY: Where a service sink is not located within a tenant space in a covered mall, the travel distance to a service sink is limited.

2021 CODE: **403.6 Service sink location.** Service sinks shall not be required to be located in individual tenant spaces in a covered mall provided that service sinks are located within a distance of travel of 300 feet (91 m) of the most remote location in the tenant space and not more than one story above or below the tenant space. Service sinks shall be located on an accessible route.

CHANGE SIGNIFICANCE: Although the code doesn't necessarily require a service sink to be located in individual tenant spaces of a covered mall, where the building arrangement requires service sinks in tenant spaces, this section allows those sinks to be located outside the tenant space, provided that they are within travel distance limitations.

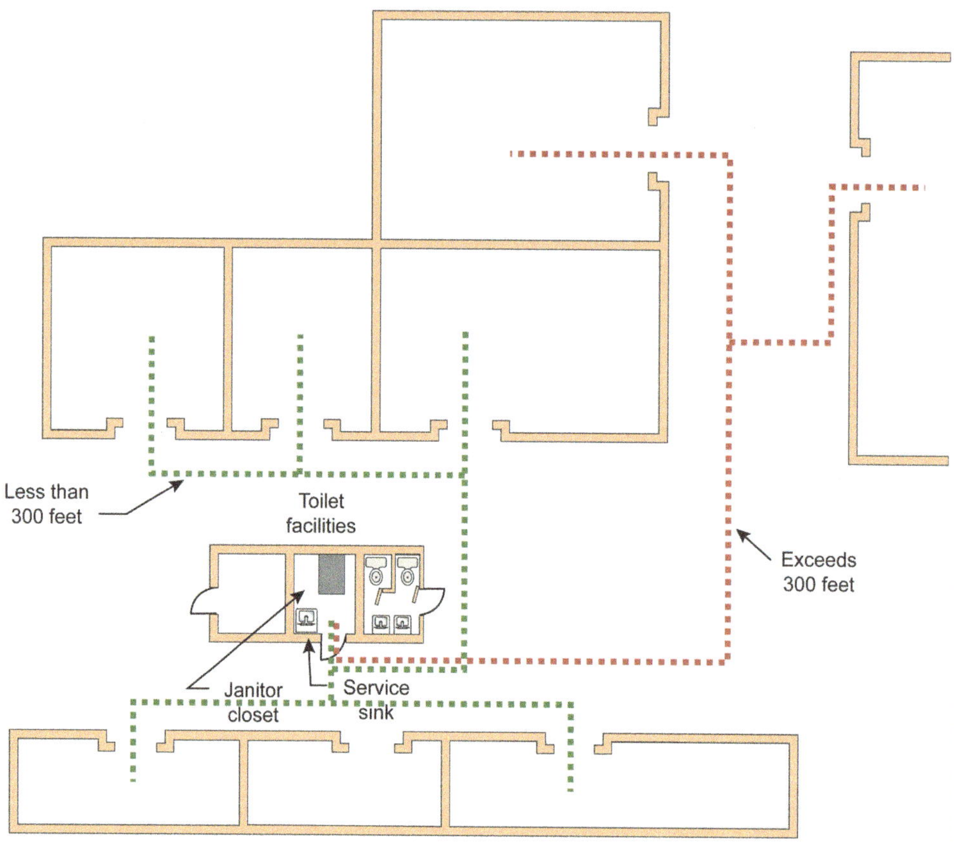

Location of service sink

International Code Council

CHANGE TYPE: Modification

CHANGE SUMMARY: A new standard for water closet carriers is added.

2021 CODE: **405.4.3 Securing wall-hung water closet bowls.** Wall-hung water closet bowls shall be supported by a concealed metal carrier that is attached to the building structural members so that strain is not transmitted to the ~~closet~~ fixture connector or any other part of the plumbing system. The carrier shall conform to ASME A112.6.1M or ASME A112.6.2.

CHANGE SIGNIFICANCE: Wall-mounted ("wall-hung") water closets require structural support within the wall. A new standard is provided for those carriers.

405.4.3

Wall Hung Fixture Carrier Standard for Water Closets

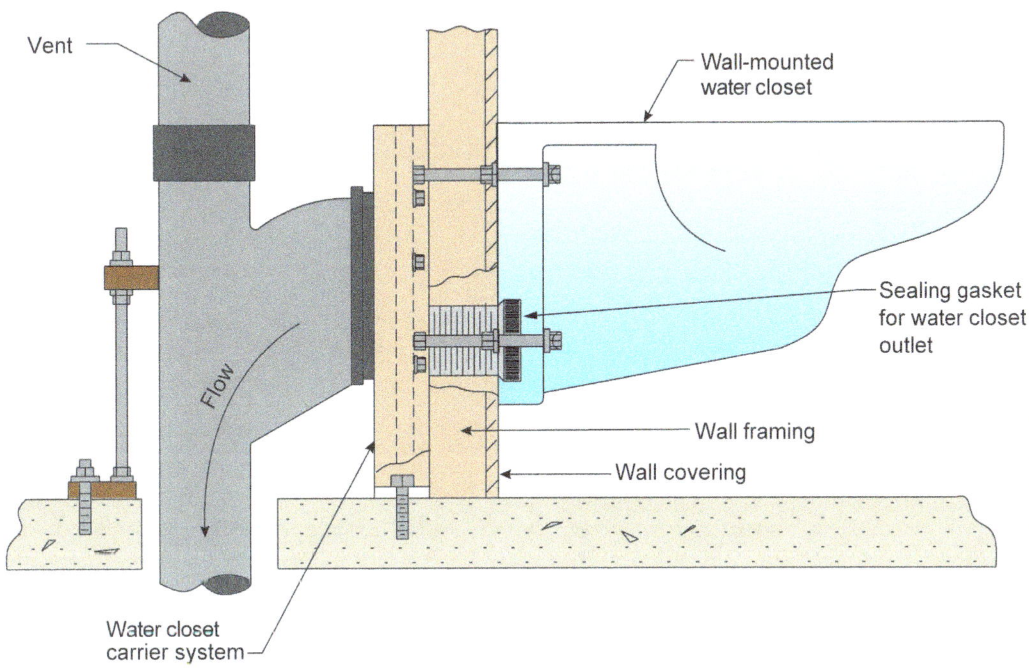

Wall-hung water closet carrier

International Code Council

407.2

Bathtubs are Not Required to have Overflow Outlets

CHANGE TYPE: Modification

CHANGE SUMMARY: A bathtub does not require an overflow outlet.

2021 CODE: **407.2 Bathtub waste outlets and overflows.** Bathtubs shall be equipped with a waste outlet ~~and an overflow outlet. The outlets shall be connected to waste tubing or piping~~ that is not less than 1½ inches (38 mm) in diameter. The waste outlet shall be equipped with a watertight stopper. Where an overflow is installed, the overflow shall be not less than 1½ inches (38 mm) in diameter.

CHANGE SIGNIFICANCE: Bathtub overflow outlets are not designed to prevent overflows of a bathtub. The flow from a tub filler faucet can easily be more flow than a typical overflow outlet can drain out of the tub. Even though an overflow outlet does allow for excess water to drain out of a tub when someone immerses themselves in a full tub (i.e., at the level of the overflow), the excess water does not instantly drain away. Protecting a tub from overflow is the bather's responsibility.

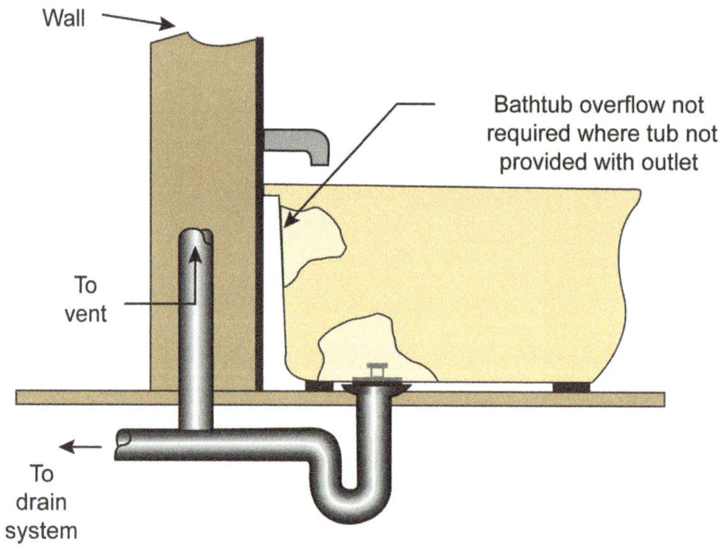

Overflow not required

International Code Council

410.3.2

Quantities of Standing versus Wheelchair Drinking Fountains

CHANGE TYPE: Addition

CHANGE SUMMARY: Fifty percent of the required number of drinking fountains must be for standing persons and the other 50 percent for persons who use wheelchairs.

2021 CODE: **410.3.2 More than the minimum number.** Where more than the minimum number of drinking fountains specified in Section 410.3.1 is provided, 50 percent of the total number of drinking fountains provided shall comply with the requirements for persons who use a wheelchair and 50 percent of the total number of drinking fountains provided shall comply with the requirements for standing persons.

Exceptions:

1. Where 50 percent of the drinking fountains yields a fraction, 50 percent shall be permitted to be rounded up or down, provided that the total number of drinking fountains complying with this section equals 100 percent of the drinking fountains.
2. Where drinking fountains are primarily for children's use, drinking fountains for people using wheelchairs shall be permitted to comply with the children's provisions in ICC A117.1 and drinking fountains for standing children shall be permitted to provide the spout at 30 inches (762 mm) minimum above the floor.

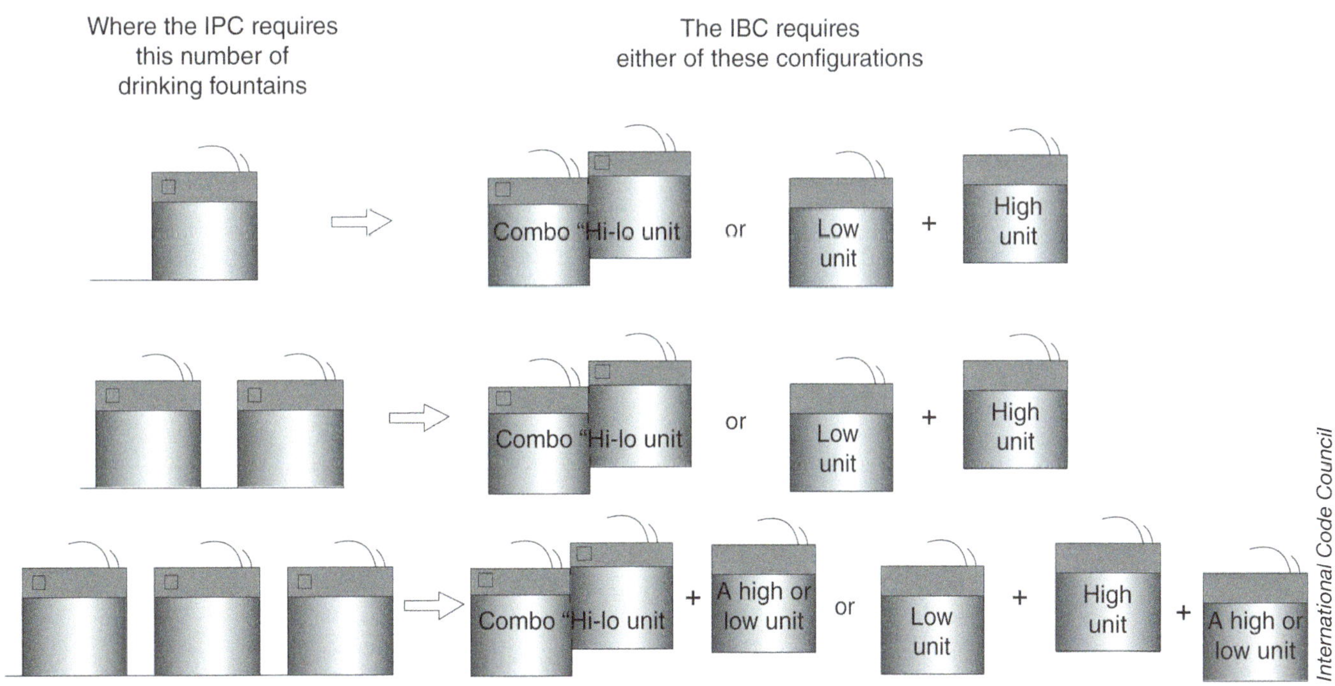

Drinking fountains for accessible design

CHANGE SIGNIFICANCE: This section has been in previous editions of the *International Building Code* and is now included in the IPC so the requirements are not overlooked. Exception 1 addresses when an odd number of drinking fountains is provided. The number can be rounded down (or up) before dividing by two. Where rounded down, the remaining required drinking fountains can be either for standing persons or persons using wheelchairs. Exception 2 applies to where children are the primary users of drinking fountains. In that case, the spout for standing children must be at least 30 inches above the floor and for children using wheelchairs, drinking fountains must comply with dimensional requirements for ICC A117.1.

410.4

Drinking Fountain Substitution using Water Dispensers

CHANGE TYPE: Clarification

CHANGE SUMMARY: At least three required drinking fountains must be installed before 50 percent of the additional required drinking fountains can be substituted with water dispensers.

2021 CODE: **410.4 Substitution.** Where restaurants provide drinking water in a container free of charge, drinking fountains shall not be required in those restaurants. In other occupancies where three or more drinking fountains are required, water dispensers shall be permitted to be substituted for not more than 50 percent of the required number of drinking fountains.

CHANGE SIGNIFICANCE: Section 410.4 is clarified to coordinate with the requirement in Section 410.3 for the minimum number of drinking fountains (for accessibility) for a building. This is not a new requirement, as Chapter 11 of the IBC requires that, at a minimum, a "high" and a "low" drinking fountain be provided. The substitution of water dispensers for drinking fountains cannot be applied to the first two drinking fountains but only to the drinking fountains that are in excess of the first two.

Photo courtesy of gettyimages.com/Spiderplay

Drinking fountain substitution using water dispensers

411.3

Water Heaters for Emergency Showers and Eye Wash Stations

CHANGE TYPE: Modification

CHANGE SUMMARY: A new type of water heater is available for emergency showers and eyewash stations that does not require a temperature-actuated mixing valve.

2021 CODE: **411.3 Water supply.** Where hot and cold water is supplied to an emergency shower or eyewash station, the temperature of the water supply shall only be controlled by a temperature-actuated mixing valve complying with ASSE 1071. Where water is supplied directly to an emergency shower or eyewash station from a water heater, the water heater shall comply with ASSE 1085.

CHANGE SIGNIFICANCE: Emergency shower stations and eyewash stations need precise limiting/control of the temperature of the water that is discharged to the user. The water discharge is also needed quickly. A water heater complying with ASSE 1085 has the necessary control technology to reliably provide water at the intended temperature for these applications.

Photo courtesy of gettyimages.com/alacatr

Emergency shower water heater

412.3

Shower Control Valves to be Rated for the Installed Shower Head

CHANGE TYPE: Modification

CHANGE SUMMARY: Lower flow shower heads need to be compatible with the shower control (mixing valve).

2021 CODE: **412.3 Individual shower valves.** Individual shower and tub-shower combination valves shall be balanced-pressure, thermostatic or combination balanced-pressure/thermostatic valves that conform to the requirements of ASSE 1016/ASME A112.1016/CSA B125.16 or ASME A112.18.1/CSA B125.1. Such valves shall be installed at the point of use. Shower control valves shall be rated for the flow rate of the installed shower head. Shower and tub-shower combination valves required by this section shall be equipped with a means to limit the maximum setting of the valve to 120°F (49°C), which shall be field adjusted in accordance with the manufacturer's instructions to provide water at a temperature not to exceed 120°F (49°C). In-line thermostatic valves shall not be utilized for compliance with this section.

412.4 Multiple (gang) showers. Multiple (gang) showers supplied with a single, tempered water supply pipe shall have the water supply for such showers controlled by an approved automatic temperature control mixing valve that conforms to ASSE 1069 or CSA B125.3, or each shower head shall be individually controlled by a balanced-pressure, thermostatic or combination balanced-pressure/thermostatic valve that conforms to ASSE 1016/ASME A112.1016/CSA B125.16 or ASME A112.18.1/CSA B125.1 and is installed at the point of use. Where a showerhead is individually controlled, shower control valves shall be rated for the flow rate of the installed shower head. Such valves shall be equipped with a means to limit the maximum setting of the valve to 120°F (49°C), which shall be field adjusted in accordance with the manufacturer's instructions to provide water at a temperature not to exceed 120°F (49°C). Access shall be provided to an ASSE 1069 or CSA B125.3 valve.

CHANGE SIGNIFICANCE: The code requires a shower head flow rate be limited to 2.5 gpm. Shower controls (mixing valves) complying with the requirements of the product standards in the code are designed to safety operate at that flow. However, there are many different types of shower heads available in the market that have lower flow rates. These lower flows might not provide enough motive power within a mixing valve to allow for the valve to self-adjust to prevent temperature shock to the user when a supply pressure or temperature change occurs. Not all shower controls (mixing valves) are designed (and tested) to accommodate lower (less than 2.5 gpm) flows.

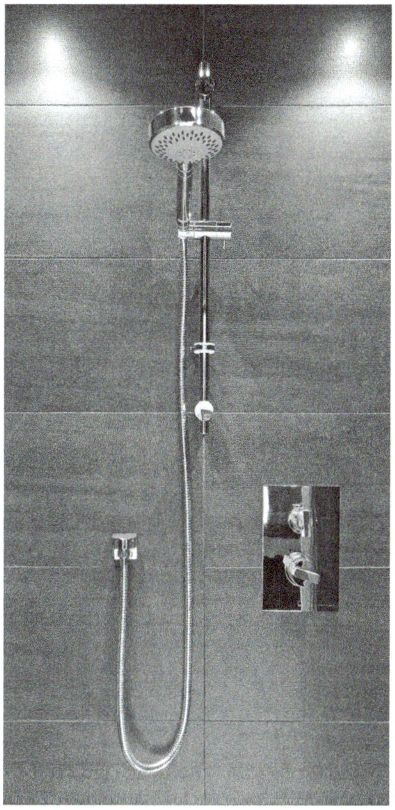

Photo courtesy of gettyimages.com/phototropic

Shower valve rated for lower flow shower head

412.5

Methods for Limiting Water Temperature Discharged to Bathtubs

CHANGE TYPE: Modification

CHANGE SUMMARY: New types of water heaters and a new design for tub faucets are additional methods that can be used to control water temperature for bathtubs.

2021 CODE: **412.5 Bathtub and whirlpool bathtub valves.** ~~The hot water supplied to~~ Bathtubs and whirlpool bathtub valves shall ~~be limited to not greater than~~ have or be supplied by a water-temperature-limiting device that conforms to ASSE 1070/ASME A112.1070/CSA B125.70 or by a water heater complying with {ASSE 1082} or {ASSE 1084}, except where ~~such protection is otherwise provided by a~~ such valves are combination tub/shower valves in accordance with Section 412.3. The water-temperature-limiting device required by this section shall be equipped with a means to limit the maximum setting of the device to 120°F (49°C), and, where adjustable, shall be field adjusted in accordance with the manufacturer's instructions to provide hot water at a temperature not to exceed 120°F (49°C). Access shall be provided to water-temperature-limiting devices that conform to ASSE 1070/ASME A112.1070/CSA B125.70.

> **Exception:** Access shall not be required for nonadjustable water-temperature-limiting devices that conform to ASSE 1070/ASME A112.1070/CSA B125.70 and are integral with a fixture fitting, provided that the fixture fitting itself can be accessed for replacement.

CHANGE SIGNIFICANCE: Two new types of water heaters are available that have the necessary control technology to reliably provide water at the intended (maximum) temperature for this application. Freestanding bathtub faucets (floor mounted) have become available that have an integral temperature limiting device that complies with the test requirement of standard ASSE 1070/ASME A112.1070/CSA B125.70. Where one of these new methods is used, an external temperature limiting device is not needed.

Photo courtesy of gettyimages.com/TriggerPhoto

Temperature limiting for bathtubs

412.10

Methods for Temperature Limitation at Head Shampoo Sinks and Footbaths

CHANGE TYPE: Modification

CHANGE SUMMARY: Two additional methods can be used to limit the water temperature discharged from a head shampoo sink.

2021 CODE: **412.10 Head shampoo sink faucets.** Head shampoo sink faucets shall be supplied with hot water that is limited to not more than 120°F (49°C)~~by a water-temperature-limiting device that conforms to ASSE 1070/ASME A112.1070/CSA B125.70~~. Each faucet shall have integral check valves to prevent crossover flow between the hot and cold water supply connections. The means for regulating the maximum temperature shall be one of the following:

1. A limiting device conforming to ASSE 1070/ASME A112.1070/CSA B125.70.
2. A water heater conforming to ASSE 1082.
3. A temperature-actuated, flow-reduction device conforming to ASSE 1062.

423.3 Footbaths and pedicure baths. The water supplied to specialty plumbing fixtures, such as pedicure chairs having an integral foot bathtub and footbaths, shall be limited to not greater than 120°F (49°C) by a water-temperature-limiting device that conforms to ASSE 1070/ASME A112.1070/CSA B125.70 or ~~CSA B125.3~~ by a water heater complying with ASSE 1082.

CHANGE SIGNIFICANCE: A water heater complying with ASSE 1082 has the necessary control technology to reliably provide water at the intended temperature for shampoo sinks, footbaths and pedicure baths. See also Section 419.5.

Photo courtesy of gettyimages.com/LightFieldStudios

Temperature limiting for head shampoo sinks

419.5

Tempered Water for Public Hand-Washing Lavatories

CHANGE TYPE: Modification

CHANGE SUMMARY: A valve conforming to CSA B125.3 is no longer acceptable as a water-temperature limiting device for public hand-washing lavatories.

2021 CODE: **419.5 Tempered water for public hand-washing facilities.** Tempered water shall be delivered from lavatories and group wash fixtures located in public toilet facilities provided for customers, patrons and visitors. Tempered water shall be delivered through an approved water-temperature limiting device that conforms to ASSE 1070/ ASME A112.1070/CSA B125.70 ~~or CSA B125.3~~.

CHANGE SIGNIFICANCE: In June of 2017, the CSA B125 Committee completed the project that removed the automatic compensating valve requirements from CSA B125.3. Therefore, valves made to the latest edition of the standard (2018) are not designed (or tested) for temperature limiting and are no longer suitable for this application.

Photo courtesy of gettyimages.com/jax10289

A CSA B125.3 device is no longer acceptable for temperature compensating.

421.3.1

Standard for Shower Waste Fittings

CHANGE TYPE: Addition

CHANGE SUMMARY: Shower drains, including linear shower drains, are required to comply with the referenced standard.

2021 CODE: **421.3.1 Waste fittings.** Waste fittings shall conform to ASME A112.18.2/CSA B125.2.

CHANGE SIGNIFICANCE: The latest edition of ASME A112.18.2/CSA B125.2 contains specific requirements for typical shower drains and also linear type drains that are different from a trench type drain. This standard includes requirements for built-up shower drain systems that are normally used in field fabricated shower systems.

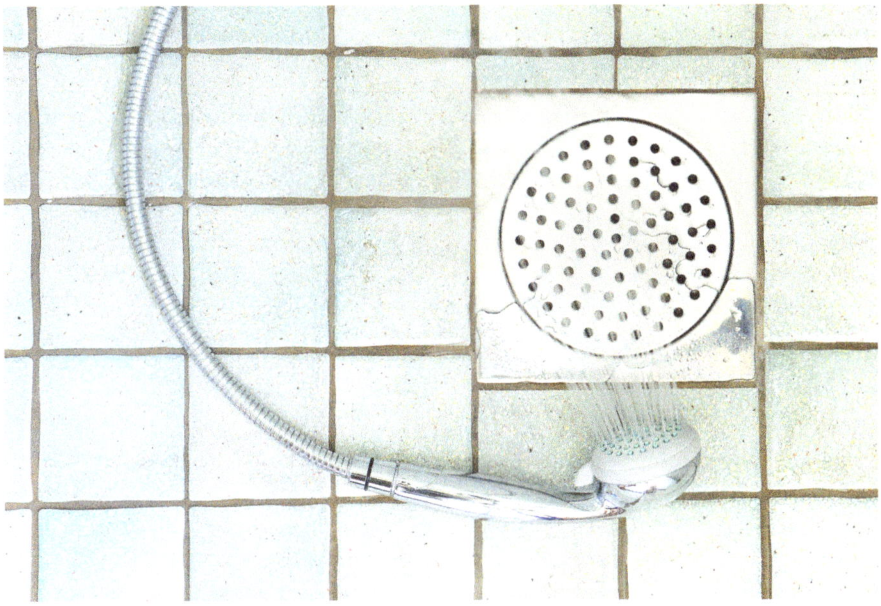

Photo courtesy of gettyimages.com/olaser

Shower drain to comply with standard

501.2

ASSE 1017 Temperature Actuated Mixing Valves

CHANGE TYPE: Clarification

CHANGE SUMMARY: The title of Standard ASSE 1017 is *Temperature-Actuated Mixing Valve*. Therefore, the code language is changed to align with the title of the standard.

2021 CODE: **501.2 Water heater as space heater.** Where a combination potable water heating and space heating system requires water for space heating at temperatures greater than 140°F (60°C), a ~~master thermostatic~~ <u>temperature-actuated</u> mixing valve complying with ASSE 1017 shall be provided to limit the water supplied to the potable hot water distribution system to a temperature of 140°F (60°C) or less. The potability of the water shall be maintained throughout the system. <u>Requirements for combination potable water heating and space heating systems shall be in accordance with the *International Mechanical Code*.</u>

CHANGE SIGNIFICANCE: The term "master" was confusing to some designers and installers as ASSE 1017 doesn't refer to that type of mixing valve. The standard indicates the acceptable location where such mixing valves are intended at the source of hot water and not at a point-of-use.

Photo courtesy of gettyimages.com/AGEphotography

"Master thermostatic" mixing valve does not exist.

602.3.5

Potable Water Pumps to Comply with NSF 61

CHANGE TYPE: Modification

CHANGE SUMMARY: Pumps used to supply drinking water must conform to NSF 61.

2021 CODE: **602.3.5 Pumps.** Pumps shall be rated for the transport of potable water. Pumps in an individual water supply system shall be constructed and installed so as to prevent contamination from entering a potable water supply through the pump units. <u>Pumps intended to supply drinking water shall conform to NSF 61.</u> Pumps shall be sealed to the well casing or covered with a water-tight seal. Pumps shall be designed to maintain a prime and installed such that ready access is provided to the pump parts of the entire assembly for repairs.

CHANGE SIGNIFICANCE: A number of major manufacturers of potable water pumps have had their pumps third-party certified to NSF 61 for many years, even though the code did not require compliance. Now that the majority of manufacturers have pumps that comply with the standard, adding this requirement to the code provides additional safety for systems that supply drinking water.

Photo courtesy of gettyimages.com/prodrive2002

NSF 61-compliant potable water pump

605.12.3, 605.13.6

Solder and Flux to Conform to NSF 61

CHANGE TYPE: Modification

CHANGE SUMMARY: Solder and flux used in making joints in pipe and tubing for drinking water systems must conform to NSF 61.

2021 CODE: **605.12.3 Solder joints.** Solder joints shall be made in accordance with ASTM B828. Cut tube ends shall be reamed to the full inside diameter of the tube end. Joint surfaces shall be cleaned. A flux conforming to ASTM B813 shall be applied. The joint shall be soldered with a solder conforming to ASTM B32. The joining of water supply piping shall be made with lead-free solder and fluxes. "Lead free" shall mean a chemical composition equal to or less than 0.2-percent lead. Solder and flux joining pipe or fittings intended to supply drinking water shall conform to NSF 61.

CHANGE SIGNIFICANCE: A number of major manufacturers of solder and flux have had their products third-party certified to NSF 61 for many years, even though the code did not require compliance. Now that the majority of manufacturers have products that comply with the standard, adding this requirement to the code provides additional safety for systems that supply drinking water.

Photo courtesy of gettyimages.com/GaryAlvis

NSF 61-Compliant solder and flux

606.1

Individual Tenant Water Shut-off Valve

CHANGE TYPE: Modification

CHANGE SUMMARY: Multiple tenant buildings must have a main water shutoff valve for each tenant space.

2021 CODE: **606.1 Location of full-open valves.** Full-open valves shall be installed in the following locations:

1. On the building water service pipe from the public water supply near the curb.
2. On the water distribution supply pipe at the entrance into the structure.
   2.1. In multiple-tenant buildings, where a common water supply piping system is installed to supply other than one- and two-family dwellings, a main shutoff valve shall be provided for each tenant.

(3. through 8. remain unchanged)

CHANGE SIGNIFICANCE: In multiple-tenant buildings, each tenant needs to have the capability to completely shut off all water coming into their space should the need arise, such as a leak in a pipe or a fixture supply valve, or where remodeling requires the modification of piping within the tenant space.

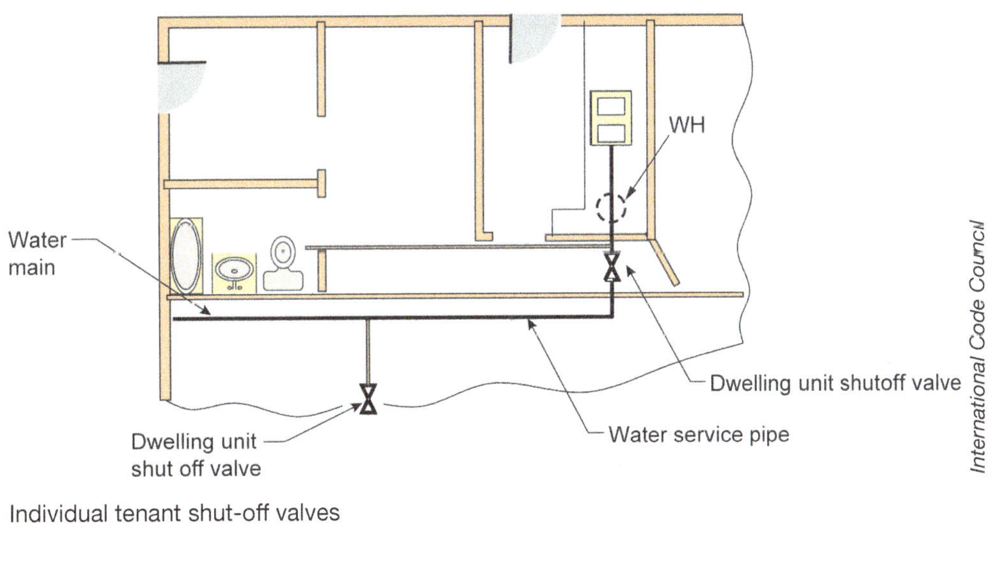

Individual tenant shut-off valves

607.1.1, 607.1.2

Water Heaters Providing Tempered Water to Fixtures

CHANGE TYPE: Modification

CHANGE SUMMARY: New designs of water heaters are available where the temperature control of the water heater can provide reliable and accurate control of the temperature of the heated water.

2021 CODE:

607.1.1 Temperature limiting means. A thermostat control for a water heater shall not only serve as the temperature limiting means for the purposes of complying with the requirements of this code for maximum allowable hot or tempered water delivery temperature at fixtures where the water heater complies with ASSE 1082 or ASSE 1085.

607.1.2 Tempered water temperature control. Tempered water shall be ~~supplied through a water temperature~~ controlled by one of the following:

1. A limiting device ~~that conforms~~ conforming to ASSE 1070/ASME A112.1070/CSA B125.70 ~~and shall limit the tempered water to not greater than~~ and set to a maximum of 110°F (43°C).
2. A thermostatic mixing valve conforming to ASSE 1017.
3. A water heater conforming to ASSE 1082.
4. A water heater conforming to ASSE 1084.

This provision shall not supersede the requirement for protective shower valves in accordance with Section 412.3.

CHANGE SIGNIFICANCE: Advances in control technologies for water heaters are recognized in the code to allow the controls of conforming water heaters to serve as the temperature limiting device rather than needing to provide an external mechanical water temperature limiting or control device.

Photo courtesy of gettyimages.com/jakkapan sapmuangphan

Water heater for tempered water

608.15.2.1

Discharge from Backflow Preventer Relief Opening

CHANGE TYPE: Modification

CHANGE SUMMARY: Where backflow preventers can relieve indoors, the relief discharge must be directed to an adequately sized waste receptor.

2021 CODE: **608.15.2.1 Relief port piping.** The termination of the piping from the relief port or air gap fitting of a backflow preventer shall discharge to an approved indirect waste receptor or to the outdoors where it will not cause damage or create a nuisance. <u>The indirect waste receptor and drainage piping shall be sized to drain the maximum discharge flow rate from the relief port as published by the backflow preventer manufacturer.</u>

CHANGE SIGNIFICANCE: The water discharge from a backflow preventer that discharges indoors must be directed to the drainage system of the building to prevent flood damage to the structure and building equipment. Alternatively, the relief discharge can be directed to the outdoors.

International Code Council

Waste receptor for backflow preventer

608.17.2

Backflow Device for Low Hazard Boiler Applications

CHANGE TYPE: Modification

CHANGE SUMMARY: The code adds a recognized standard, ASSE 1081, for a combination pressure regulator/backflow preventer product for boilers.

2021 CODE: **608.17.2 Connections to boilers.** The potable supply to the boiler shall be equipped with a backflow preventer with an intermediate atmospheric vent complying with ASSE 1012, ASSE 1081, or CSA B64.3. Where conditioning chemicals are introduced into the system, the potable water connection shall be protected by an air gap or a reduced pressure principle backflow preventer, complying with ASSE 1013, CSA B64.4 or AWWA C511.

CHANGE SIGNIFICANCE: Potable water connections to boilers require both a pressure reducing valve and a backflow prevention device. The added standard, ASSE 1081, *Performance Requirements for Backflow Preventers with Integral Pressure Reducing Boiler Feed Valve and Intermediate Atmospheric Vent Style for Domestic and Light Commercial Water Distribution Systems*, recognizes dual-purpose devices that satisfy both requirements. Products are now available that are "all-in-one" to simplify installations for low hazard boiler connections.

Photo courtesy of gettyimages.com/Yliya Zhuravleva

Pressure reducing valve/backflow preventer for boiler

609.2

Two Water Service Pipes for Group I-2, Condition 2 Healthcare Facilities

CHANGE TYPE: Modification

CHANGE SUMMARY: Buildings classified as Group I-2, Condition 2 facilities require two water service pipes. The previous requirement for hospitals to have two water service pipes was vague, resulting in enforcement difficulties.

2021 CODE: **609.2 Water service for Group I-2, Condition 2 facilities.** ~~Hospitals~~ Group I-2, Condition 2 facilities shall have not fewer than two water service pipes ~~installed in such a manner so as to minimize the potential for an interruption of the supply of water in the event of a water main or water service pipe failure.~~ sized such that with the loss of the largest service pipe, the remaining service pipes will meet the water demand for the entire facility. Each water service shall have a shutoff valve in the building and a shutoff valve at the utility-provided point of connection to the water main or other source of potable water.

CHANGE SIGNIFICANCE: Determining which hospitals actually needed redundant water service pipes could only be based on whether the facility was actually named a "hospital." Many facilities that do require water service redundancy may not be named "hospital." The use of the terminology "Group I-2, Condition 2 facilities," as detailed in Chapter 3 of the IBC provides clarity as to which buildings require the redundancy based on occupancy classification. The modification adds requirements for shutoff valves on those water lines and recognizes that not all such facilities will be served by a water utility (some could have ground water wells or other means.)

Photo courtesy of gettyimages.com/GoodLifeStudio

Healthcare facility water service pipes

609.2.1

Tracer Wire for Buried Nonmetallic Water Service Piping

CHANGE TYPE: Addition

CHANGE SUMMARY: The addition of a tracer wire on buried hospital water service piping allows for easier locating to avoid piping damage that would disrupt water service.

2021 CODE: **609.2.1 Tracer wire for nonmetallic piping.** An insulated tracer wire listed for the purpose or other approved conductor shall be installed adjacent to underground nonmetallic piping serving as a water service for a hospital. Access shall be provided to the tracer wire or the tracer wire shall terminate above ground at each end of the nonmetallic piping. The tracer wire size shall be not less than 18 AWG and the wire insulation type shall be suitable for direct burial.

CHANGE SIGNIFICANCE: A tracer wire is now required for underground nonmetallic water service piping serving Group I-2, Condition 2 facilities (hospitals). Although the installation of tracer products for buried nonmetallic piping and conduit is common in the utility industry, this section is specific as to the minimum size of the wire, access to the wire at each end of the piping and the insulation on the wire needing to be listed for direct burial service. Typical building wiring is not likely to be listed (by a third-party certification agency) for tracer wire service. Note that as AWG wire size number designations become larger, the wire diameters become smaller. Sizes less than 18 AWG are products having smaller wire diameters.

Photo courtesy of gettyimages.com/Supersmario

Tracer required for underground nonmetallic water service pipe

Table 702.3

ABS Building Sewer Pipe Standard

CHANGE TYPE: Modification

CHANGE SUMMARY: A standard is added to the building sewer pipe table for a composite wall ABS pipe.

2021 CODE:

TABLE 702.3 Building Sewer Pipe

Material	Standard
Acrylonitrile butadiene styrene (ABS) plastic pipe in IPS diameters, including Schedule 40, DR 22 (PS 200) and DR 24 (PS 140); with a solid, cellular core or composite wall	ASTM D2661; ASTM D2680; ASTM F628; ASTM F1488; CSA B181.1

CHANGE SIGNIFICANCE: Another form of ABS pipe is added to the table of building sewer pipe standards. The standard covers nominal pipe sizes of 8, 10, 12 and 15 inches. This form of pipe is a composite wall construction where concentric inner and outer walls are braced by an integral truss-type structure. The resultant annular space between the walls is filled with a Portland cement/pearlite mixture. The product has a pipe stiffness of at least 200 psi to provide for very low deflection values for deep burial conditions as well as for AASHTO live loading conditions at less than one foot of cover.

Photo courtesy of gettyimages.com/nehrut

New standard for ABS sewer pipe

705.2.4, 705.10.4

PVC and ABS Push-fit DWV Fittings

CHANGE TYPE: Addition

CHANGE SUMMARY: Push-fit fittings are a new type of DWV fitting for ABS and PVC piping that, when used, will reduce installation time.

2021 CODE: **705.2.4 Push-fit joints.** Push-fit DWV fittings shall be listed and labeled to ASME A112.4.4 and shall be installed in accordance with the manufacturer's instructions.

705.10.4 Push-fit joints. Push-fit joints shall conform to ASME A112.4.4 and shall be installed in accordance with the manufacturer's instructions.

CHANGE SIGNIFICANCE: Push-fit DWV fittings of either ABS or PVC are available for use. Such fittings eliminate the need for the solvent cementing of joints. The new standard has been added to Table 702.4, Pipe Fittings.

Photo courtesy of gettyimages.com/DenBoma

Push-fit drainage fittings

708.1.6

Removable Fixture Traps Serving as Cleanouts

CHANGE TYPE: Addition

CHANGE SUMMARY: Removable traps and removable fixtures with integral traps are acceptable as equivalent to cleanouts.

2021 CODE: **708.1.6 Cleanout equivalent.** A fixture trap or a fixture with integral trap, removable without altering concealed piping, shall be acceptable as a cleanout equivalent.

CHANGE SIGNIFICANCE: Removable traps and removable fixtures with integral traps such as urinals and water closets have long been considered as acceptable access to the drainage system for clearing stoppages. Note that the current code already has a few limitations on these cleanout means: 1) Section 708.1.3 prohibits the removal of water closets to serve as cleanout access to a building sewer. 2) Section 708.1.5 (Exception 1) only allows cleanout access through a removable P-trap for the same or one size larger pipe size.

Photo courtesy of gettyimages.com/ClausAlwinVogel

Removable trap as cleanout

717, 718

Methods for Restoring Building Sewer Piping

CHANGE TYPE: Addition

CHANGE SUMMARY: The code recognizes two methods for restoring building sewer and building drain piping.

2021 CODE:

SECTION 717
RELINING BUILDING SEWERS AND BUILDING DRAINS

717.1 General. This section shall govern the relining of existing building sewers and building drainage piping.

717.2 Applicability. The relining of existing building sewers and building drainage piping shall be limited to gravity drainage piping 4 inches (102 mm) in diameter and larger. The relined piping shall be of the same nominal size as the existing piping.

717.3 Pre-installation requirements. Prior to commencement of the relining installation, the existing piping sections to be relined shall be descaled and cleaned. After the cleaning process has occurred and water has been flushed through the system, the piping shall be inspected internally by a recorded video camera survey.

717.3.1 Pre-installation recorded video camera survey. The video survey shall include verification of the project address location. The video shall include notations of the cleanout and fitting locations, and the approximate depth of the existing piping. The video shall also include notations of the length of piping at intervals not greater than 25 feet (7620 mm).

717.4 Permitting. Prior to permit issuance, the code official shall review and evaluate the preinstallation recorded video camera survey to determine if the piping system is able to be relined in accordance with the proposed lining system manufacturer's installation requirements and applicable referenced standards.

717.5 Prohibited applications. Where review of the preinstallation recorded video camera survey reveals that piping systems are not installed correctly or defects exist, relining shall not be permitted. The defective portions of piping shall be exposed and repaired with pipe and fittings in accordance with this code. Defects shall include, but are not limited to, backgrade or insufficient slope, complete pipe wall deterioration or complete separations such as from tree root invasion or improper support.

717.6 Relining materials. The relining materials shall be manufactured in compliance with applicable standards and certified as required in Section 303. Fold-and-form pipe reline materials shall be manufactured in compliance with ASTM F1504 or ASTM F1871.

717.7 Installation. The installation of relining materials shall be performed in accordance with the manufacturer's installation instructions, applicable referenced standards and this code.

717.7.1 Material data report. The installer shall record the data as required by the relining material manufacture and applicable standards. The recorded data shall include but is not limited to the location of the project, relining material type, amount of product installed and conditions of the installation. A copy of the data report shall be provided to the code official prior to final approval.

717.8 Post-installation recorded video camera survey. The completed relined piping system shall be inspected internally by a recorded video camera survey after the system has been flushed and flow-tested with water. The video survey shall be submitted to the code official prior to finalization of the permit. The video survey shall be reviewed and evaluated to provide verification that no defects exist. Any defects identified shall be repaired and replaced in accordance with this code.

717.9 Certification. A certification shall be provided in writing to the code official, from the permit holder, that the relining materials have been installed in accordance with the manufacturer's installation instructions, the applicable standards and this code.

717.10 Approval. Upon verification of compliance with the requirements of Sections 717.1 through 717.9, the code official shall approve the installation.

Building sewer restoration

SECTION 718
REHABILITATION OF BUILDING SEWERS AND BUILDING DRAINS

718.1 Cure-in-place. Sectional cure-in-place rehabilitation of building sewer piping and sewer service lateral piping shall be in accordance with ASTM F2599. Main and lateral cure-in-place rehabilitation of building sewer and sewer service lateral pipe and their connections to the main sewer pipe shall be in accordance with ASTM F2561. Hydrophilic rings or gaskets in cure-in-place rehabilitation of building sewer piping and sewer service laterals shall be in accordance with ASTM F3240 to ensure water tightness and elimination of ground water penetration.

CHANGE SIGNIFICANCE: The methods covered in Sections 717 and 718 are considered as trenchless pipe restoration technologies. Section 717 covers a process referred to as the "fold and form pipe (FFP)" method of pipe restoration. A factory-made extruded thermoplastic folded liner is pulled into a cleaned existing pipe, expanded with air and steam (steam for softening the liner to allow unfolding) and then allowed to cool (using air pressure only) to form a new pipe within the existing pipe.

Section 718 covers a process referred to as the "cured-in-place pipe (CIPP)" method of pipe restoration. A felt tube is inserted into a flexible calibration tube (for ensuring the final cured pipe size.) The felt tube is impregnated in the field with polyester or epoxy resin. The tube assembly is inverted as it is inserted (or pulled) into the cleaned existing pipe so as to place the calibration tube on the inside of the felt liner. The calibration tube is expanded, and various methods are used to cure the resin in the felt liner. The calibration tube is removed after curing is complete.

903.1.3

Protected Outdoor Vent Termination Method

CHANGE TYPE: Addition

CHANGE SUMMARY: New requirements for protected outdoor roof vent terminals accommodate solar panel and architectural roof feature installations.

2021 CODE: **903.1.3 Protected vent terminal.** Where an open vent pipe terminates above a sloped roof and is covered by either a roof-mounted panel (such as a solar collector or photovoltaic panel mounted over the vent opening) or a roof element (such as an architectural feature or a decorative shroud), the vent pipe shall terminate not less than 2 inches (51 mm) above the roof surface. Such roof elements shall be designed to prevent the adverse effects of snow accumulation and wind on the function of the vent. The placement of a panel over a vent pipe and the design of a roof element covering the vent pipe shall provide for an open area for the vent pipe to the outdoors that is not less than the area of the pipe, as calculated from the inside diameter of the pipe. Such vent terminals shall be protected by a method that prevents birds and rodents from entering or blocking the vent pipe opening.

CHANGE SIGNIFICANCE: There is an increasing trend towards buildings having rooftop-mounted solar panel arrays that is often at odds with the location of plumbing vent terminals on the roof. Also, the architectural community sometimes wishes to obscure vent piping terminations above the roof from view. Although vent piping systems could be designed to cause the vent terminals to be located elsewhere, doing so isn't necessarily cost-effective, especially where work on existing buildings is concerned. This new section provides details for having roof vent terminals covered (protected) by either solar panels or architectural shrouds.

Photo courtesy of gettyimages.com/bigdaddyrockstar

Architectural shroud

915.1

Food Waste Disposers on Combination Waste and Vent Systems

CHANGE TYPE: Modification

CHANGE SUMMARY: The prohibition of a food waste disposer discharging to a combination waste and vent system is removed.

2021 CODE: **915.1 Type of fixtures.** A combination waste and vent system shall not serve fixtures other than floor drains, sinks, lavatories and drinking fountains. Combination waste and vent systems shall not receive the discharge from a ~~food waste disposer or~~ clinical sink.

CHANGE SIGNIFICANCE: Laboratory investigations and years of recent field experience have proven that the prohibition on combination waste and vent systems receiving the discharge of food waste disposers is unwarranted. This change will allow for more applications of combination waste and vent systems in kitchen applications.

Photo courtesy of gettyimages.com/JackValley

Food waste disposer on CWV system

1002.1

Direct Connection to Hydromechanical Grease Interceptor

CHANGE TYPE: Modification

CHANGE SUMMARY: A one, two or three-compartment pots and pans sink without a trap can be directly connected to a hydromechanical grease interceptor provided that the grease interceptor is in close proximity and connects to a drainage branch that has an emergency floor drain connected immediately downstream of the interceptor.

2021 CODE: **1002.1 Fixture traps.** Each plumbing fixture shall be separately trapped by a liquid-seal trap, except as otherwise permitted by this code. The vertical distance from the fixture outlet to the trap weir shall not exceed 24 inches (610 mm), and the horizontal distance shall not exceed 30 inches (610 mm) measured from the centerline of the fixture outlet to the centerline of the inlet of the trap. The height of a clothes washer standpipe above a trap shall conform to Section 802.3.3. A fixture shall not be double trapped.

Exceptions:

1. and **2.** *(remain unchanged)*
3. ~~A grease interceptor intended to serve as a fixture trap in accordance with the manufacturer's installation instructions shall be permitted to serve as the trap for a single fixture or a combination sink of not more than three compartments where the vertical distance from the fixture outlet to the inlet of the interceptor does not exceed 30 inches (762 mm) and the developed length of the waste pipe from the most upstream fixture outlet to the inlet of the interceptor does not exceed 60 inches (1524 mm).~~

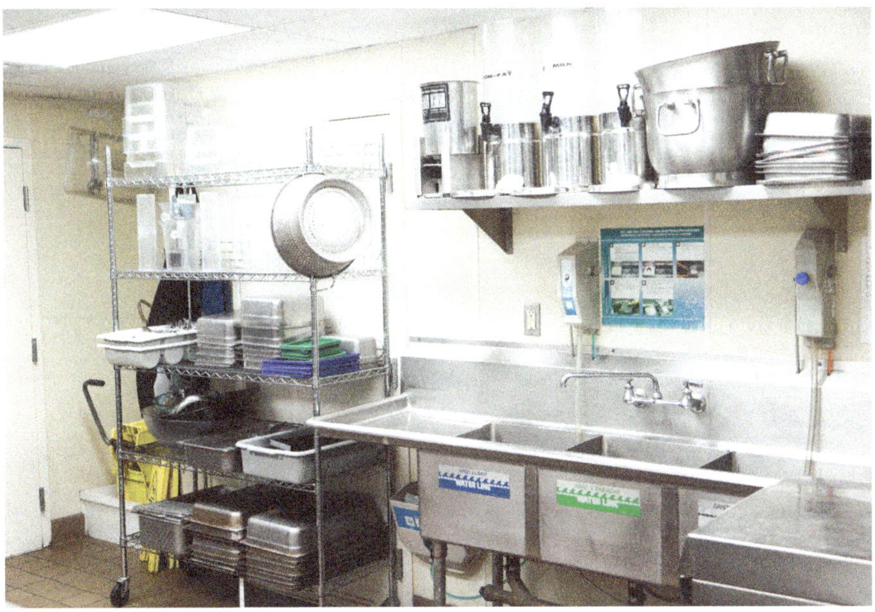

Photo courtesy of gettyimages.com/slobo

Direct connection to hydromechanical interceptor permitted

~~4.~~ 3. Floor drains in multilevel parking structures that discharge to a building storm sewer shall not be required to be individually trapped. Where floor drains in multilevel parking structures are required to discharge to a combined building sewer system, the floor drains shall not be required to be individually trapped provided that they are connected to a main trap in accordance with Section 1103.1.

4. Where a hydromechanical grease interceptor serves a food utensil, dishes, pots-and-pans sink, in accordance with the manufacturer's installation instructions. The branch drain serving the interceptor shall be provided with an emergency floor drain downstream of the interceptor connection, and the branch shall serve only the emergency floor drain and the interceptor. Where the interceptor serves a combination sink of not more than three compartments where the vertical distance from the fixture outlet to the inlet of the interceptor does not exceed 30 inches (762 mm) and the developed length of the waste pipe from the most upstream fixture outlet to the inlet of the interceptor does not exceed 60 inches (1524 mm). The food utensil, dishes, pots and pans sink shall be required to connect directly with the interceptor.

CHANGE SIGNIFICANCE: Section 802.1.7 requires pots and pans sinks to be indirectly connected to the drainage system to prevent drainage system backups from contaminating the sink wells. An indirect connection allows a fixture to not have a trap (i.e., the fixture falls under the "except as otherwise permitted by this code" of the first sentence of Section 1002.1). Exception 4, in effect, overrides the indirect connection requirement of Section 802.1.7 for pots and pans sinks but also allows that direct connection to not have a trap where a hydromechanical grease interceptor manufacturer's installation instructions indicate that a fixture trap is not required, and the hydromechanical grease interceptor is connected to the drainage system in a specific manner. In other words, the hydromechanical grease interceptor becomes an extension of the waste piping from the pots and pans sink provided that the interceptor is connected to the drainage system in a manner where if the drainage system did back up, the backup would only overflow at the emergency flow drain and not into the pots and pans sink.

1002.4.1.5

Fixture Drains Serving as a Trap Priming Method

CHANGE TYPE: Addition

CHANGE SUMMARY: Waste from lavatories and hand sinks can be directed to floor drains, trench drains or floor sinks where such floor fixtures require a trap primer.

2021 CODE: **1002.4.1.5 Fixture drain connection for trap priming.** A fixture drain from a lavatory or hand sink shall serve as a method of providing trap seal protection for an emergency floor drain, a trench drain, or a floor sink where such fixtures are located in the same room. A fixture drain from a drinking fountain shall serve as a method of providing trap seal protection for an emergency floor drain, a trench drain, or a floor sink where such fixtures are in the same room or in a room adjacent to the room having the drinking fountain. The fixture drain shall not be routed on or above the surface of the floor and shall connect to the floor drain, trench drain, or floor sink at a point that is below the flood level rim and above the inlet to the trap of the receiving fixture.

CHANGE SIGNIFICANCE: This method is an extension of the concept in Section 1002.4.1.3 where a portion of the waste flow from a lavatory is directed to another fixture requiring its trap seal to be protected from evaporation. This method allows the entire flow to be directed to such traps.

Photo courtesy of gettyimages.com/ Navebreakmedia

Fixture drain permitted to serve as trap priming method

1102.6

Roof Drains to be Tested and Rated for Flow

CHANGE TYPE: Modification

CHANGE SUMMARY: Flow testing and rating of roof drains is required to be in accordance with the referenced standards.

2021 CODE: **1102.6 Roof drains.** Roof drains shall conform to ASME A112.3.1 or ASME A112.6.4. Roof drains, other than siphonic roof drains, shall be tested and rated in accordance with ASME A112.6.4 or ASPE/IAPMO Z1034.

CHANGE SIGNIFICANCE: For the 2021 IPC, Section 1106.2 requires that sizing of the chosen storm drain piping be checked against the published discharge flow rate of the specific roof drain model chosen for the storm drain pipe. Requiring testing and rating of roof drains to standards will result in third party certified flow rate information that the roof drain manufacturers can use for publishing flow rate data.

Photo courtesy of gettyimages.com/DenBoma

Product testing and rating for flow required

1106.2.1

Rainfall Rate Conversion Method

CHANGE TYPE: Addition

CHANGE SUMMARY: Although the conversion from inches-per-hour of rainfall to gallons per minute (gpm) is not complicated, including the conversion information in the code is helpful.

2021 CODE: **1106.2.1 Rainfall rate conversion method.** The rainfall rate falling on a roof surface shall be converted to a gallon per minute (L/m) flow rate in accordance with Equation 11-1.

$$GPM = R \times A \times 0.0104 \quad \text{(Equation 11-1)}$$

where:

R = Rainfall intensity in inches (mm) per hour.

A = Roof area in square feet (m^2).

CHANGE SIGNIFICANCE: Because the code requires sizing of the storm drainage piping using gallons per minute flow, having an equation to make the conversion of rainfall rate into gallons per minute clarifies the calculation process.

Photo courtesy of gettyimages.com/Willowpix

A new equation converts inches per hour of rainfall to gallons per minute flow rate.

1202.1

Nonflammable Medical Gas Systems

CHANGE TYPE: Modification

CHANGE SUMMARY: NFPA 99 covers the installation, testing and labels for nonflammable medical gases.

2021 CODE: **1202.1 Nonflammable medical gases.** Nonflammable medical gas systems, inhalation anesthetic systems and vacuum piping systems shall be ~~designed~~ installed, tested and ~~installed~~ labeled in accordance with NFPA 99.

Exceptions:

1. This section shall not apply to portable systems or cylinder storage.
2. Vacuum system exhaust terminations shall comply with the *International Mechanical Code*.

CHANGE SIGNIFICANCE: System designers are responsible for the design of nonflammable medical gas systems, inhalation anesthetic systems and vacuum piping systems. The installation, testing and labels for these systems must be in accordance with the referenced NFPA standard.

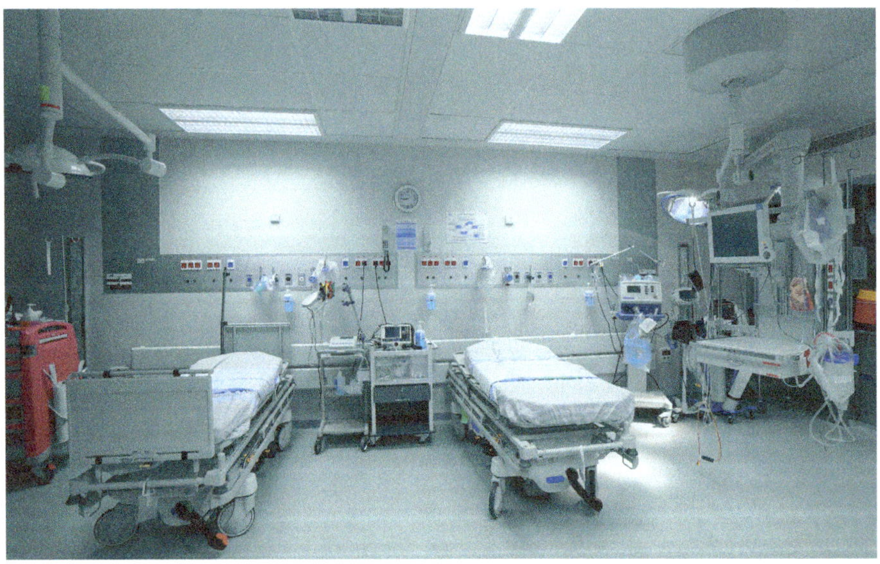

Photo courtesy of gettyimages.com/alexey_ds

Nonflammable medical gas system installation

1301.1

Nonpotable Rainwater Standard Alternative

CHANGE TYPE: Modification

CHANGE SUMMARY: Standard CSA B805/ICC 805 is added to the code to serve as an alternative method for providing a nonpotable water source.

2021 CODE: **1301.1 ~~Scope~~ General.** The provisions of Chapter 13 shall govern the materials, design, construction and installation of systems for the collection, storage, treatment and distribution of nonpotable water. For nonpotable rainwater systems, the provisions of CSA B805/ICC 805 shall be an alternative for regulating the materials, design, construction and installation of systems for rainwater collection, storage, treatment and distribution of nonpotable water. The use and application of nonpotable water shall comply with laws, rules and ordinances applicable in the jurisdiction.

CHANGE SIGNIFICANCE: Although Chapter 13 offers good overall coverage for using rainwater as a nonpotable water source, CSA B805/ICC 805, *Rainwater Harvesting Systems*, offers a more comprehensive approach for these systems.

Photo courtesy of gettyimages.com/sherjaca

Rainwater storage system

SIGNIFICANT CHANGES TO THE

INTERNATIONAL MECHANICAL CODE®

2021 EDITION

PART 2

International Mechanical Code

Chapters 1 through 15

- **Chapter 1** Scope and Administration
  No changes addressed
- **Chapter 2** Definitions
  No changes addressed
- **Chapter 3** General Regulations
- **Chapter 4** Ventilation
- **Chapter 5** Exhaust Systems
- **Chapter 6** Duct Systems
- **Chapter 7** Combustion Air
  No changes addressed
- **Chapter 8** Chimneys and Vents
- **Chapter 9** Specific Appliances, Fireplaces and Solid Fuel-Burning Equipment
- **Chapter 10** Boilers, Water Heaters and Pressure Vessels
  No changes addressed
- **Chapter 11** Refrigeration
- **Chapter 12** Hydronic Piping
  No changes addressed
- **Chapter 13** Fuel Oil Piping and Storage
  No changes addressed
- **Chapter 14** Solar Thermal Systems
  No changes addressed
- **Chapter 15** Referenced Standards
  No changes addressed

The *International Mechanical Code*® (IMC®) contains provisions for the regulation of mechanical equipment design and installation. The code consists of fifteen chapters and two appendices. Appendix A deals with chimney connector pass-throughs, and Appendix B lists recommended permit fees.

The provisions of Chapter 1 address the scope, application, enforcement and administration of subsequent requirements of the code. Chapter 2 provides definitions for terms used throughout the IMC. Chapter 3 includes the general requirements for listed equipment, appliance location, protection for personnel servicing mechanical equipment, access requirements for appliances in various locations and condensate disposal. Chapter 4 addresses building ventilation. Chapter 5 addresses exhaust systems, including commercial kitchen exhaust systems. Chapter 6 covers all of the duct construction provisions for heating, ventilation and air conditioning systems. Chapter 7 addresses combustion air. Chapter 8 has requirements for vents and chimneys. Chapter 9 includes requirements for specific appliances, fireplaces and solid-fuel equipment. Chapter 10 deals with boilers, water heaters and pressure vessels. Chapter 11 addresses refrigeration systems. Chapter 12 has requirements for hydronic piping. Regulations governing fuel-oil piping and storage are located in Chapter 13. Chapter 14 deals with solar thermal systems. Chapter 15 contains standards, referenced in the 2018 IMC, listed by the promulgating agency of the standard. ■

307.2.1.1

Condensate Discharge

401.2, 403.1

Mechanical Ventilation for Dwelling Units

401.4, 501.3.1

Intake Opening and Exhaust Outlet Locations

TABLE 403.3.1.1

Recirculation of Mechanical Exhaust Prohibited

403.3.1.3

Demand Controlled Ventilation

403.3.2.1

Outdoor Air for Dwelling Units

502.20

Manicure and Pedicure Station Exhaust System

504.4.1

Termination Location for Dryer Exhaust

506.3.7

Factory-Built Grease Duct Slope

507.1

Smoker Ovens with Integral Exhaust

514.2

Energy Recovery Ventilation Systems

602.2.1.8

Pipe and Duct Insulation Within Plenums

607.5.2

Duct Penetrations of Fire Barriers

607.5.5

Subducts Penetrating Shaft Enclosures

608.1, 403.3.1.5

Ventilation Air Distribution

801.21

Blocked Vent Switch for Oil-Fired Appliances

905.1

Wood-Burning Residential Hydronic Heaters

929

Unvented Alcohol Fuel-Burning Decorative Appliances

1101 THROUGH 1108

Ammonia Refrigeration Systems

1105.9

Machinery Room Means of Egress

1107 THROUGH 1110

Refrigerant Piping

307.2.1.1

Condensate Discharge

CHANGE TYPE: Modification

CHANGE SUMMARY: Significant coverage is added to the code regarding the point of disposal of condensate. Such coverage addresses what has been common practice (acceptable and unacceptable) in most jurisdictions.

2021 CODE: **307.2.1.1 Condensate discharge.** Condensate drains shall not directly connect to any plumbing drain, waste or vent pipe. Condensate drains shall not discharge into a plumbing fixture other than a floor sink, floor drain, trench drain, mop sink, hub drain, standpipe, utility sink or laundry sink. Condensate drain connections to a lavatory wye branch tailpiece or to a bathtub overflow pipe shall not be considered as discharging to a plumbing fixture. Except where discharging to grade outdoors, the point of discharge of condensate drains shall be located within the same occupancy, tenant space or dwelling unit as the source of the condensate.

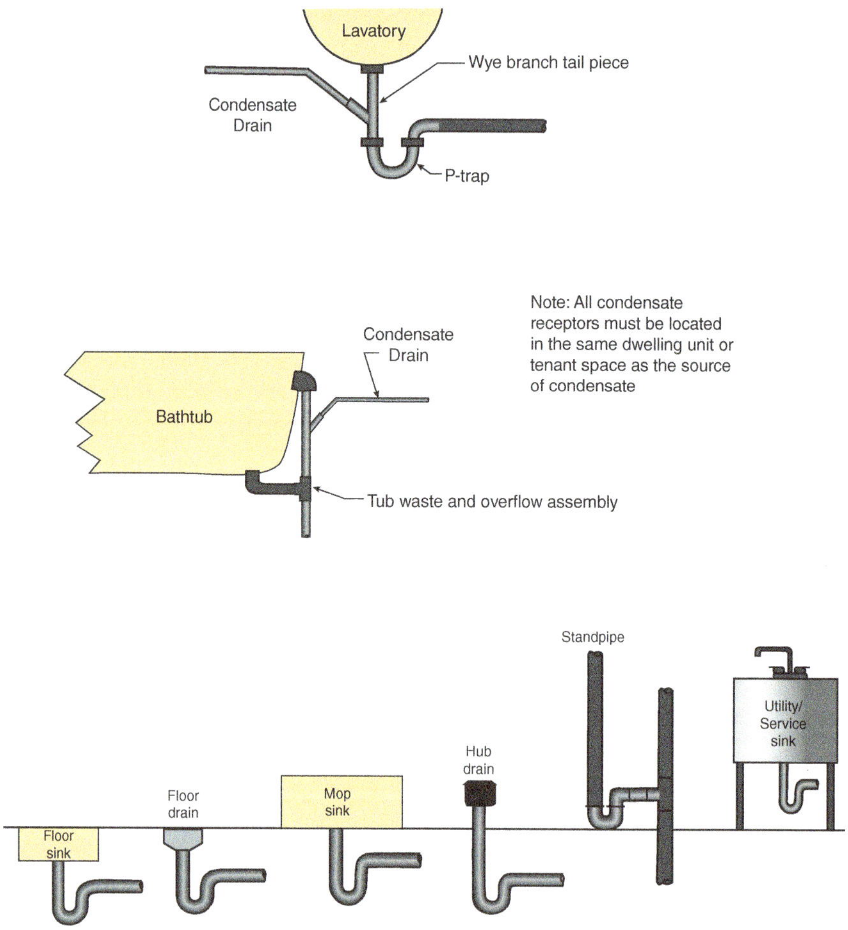

International Code Council

Allowable receptors for condensate

CHANGE SIGNIFICANCE: This new section states prohibitions on what are bad practices and also states allowances for what are considered to be acceptable practices for condensate discharge. The code was silent on these practices until now. This text expressly prohibits the connection of condensate drains directly to DWV piping and prohibits condensate from discharging into plumbing fixtures other than those listed. The intent is to prevent insanitary conditions and potential health hazards.

Health hazards are created by drilling holes in plumbing vent piping in attics to connect a condensate drain from HVAC appliances installed in the attic. This is but one example of the hazards this new text intends to prevent. This text also clarifies that connections to lavatory tailpieces and bathtub overflow pipes are acceptable, as this has been a common practice in some localities.

401.2, 403.1

Mechanical Ventilation for Dwelling Units

CHANGE TYPE: Clarification

CHANGE SUMMARY: This change clarifies where mechanical ventilation is required for dwelling units. The previous reference to the testing required by Section R402.4.1.2 of the *International Energy Conservation Code* (IECC) had caused varying interpretations as to whether the testing applied only to R-2 buildings of 3 stories or less or if it applied to all R-2 buildings. The revised text makes no distinction between R-2 buildings of 3 stories or less and R-2 buildings over 3 stories, thereby eliminating multiple interpretations.

2021 CODE: 401.2 Ventilation required. Every occupied space shall be ventilated by natural means in accordance with Section 402 or by mechanical means in accordance with Section 403. ~~Where the air infiltration rate in a dwelling unit is less than 5 air changes per hour when tested with a blower door at a pressure of 0.2-inch water column (50 Pa) in accordance with Section R402.4.1.2 of the *International Energy Conservation Code*, the dwelling unit shall be ventilated by~~ Dwelling units complying with the air leakage requirements of the *International Energy Conservation Code* or *ASHRAE 90.1* shall be ventilated by mechanical means in accordance with Section 403. Ambulatory care facilities and Group I-2 occupancies shall be ventilated by mechanical means in accordance with Section 407.

403.1 Ventilation system. Mechanical ventilation shall be provided by a method of supply air and return or exhaust air except that mechanical ventilation air requirements for Group R-2, R-3 and R-4 occupancies ~~three stories and less in height above grade plane~~ shall be provided by an exhaust system, supply system or combination thereof. The amount of supply air shall be approximately equal to the amount of return and exhaust air. The system shall not be prohibited from producing negative or positive pressure. The system to convey ventilation air shall be designed and installed in accordance with Chapter 6.

Photo courtesy of gettyimages.com/Lana2011

Group R-2 apartment building

CHANGE SIGNIFICANCE: The requirement for mechanical ventilation in R-2 dwelling units is no longer tied to a residential blower door testing requirement. This eliminates the distinction between commercial and residential R-2 buildings as defined in the IECC. The IMC does not regulate one- and two-family dwellings and townhouses, but does regulate R-2 multiple-family buildings, and the intent of this code section is now evident regarding R-2 buildings, that is, dwelling units under the scope of the IMC must be mechanically ventilated if such units comply with the air leakage requirements of the applicable energy code. In other words, if the building is constructed to significantly limit air leakage through the thermal envelope, then the building will have few air changes per hour and mechanical ventilation will be necessary to provide a healthy environment in the dwelling units. Section 403.1 was revised for consistency, as the number of stories above grade is no longer relevant.

401.4, 501.3.1

Intake Opening and Exhaust Outlet Locations

CHANGE TYPE: Modification

CHANGE SUMMARY: A new type of factory-built combination exhaust and intake air fitting is introduced that does not require separation between the two openings.

2021 CODE: **401.4 Intake opening location.** Air intake openings shall comply with all of the following:

1. Intake openings shall be located not less than 10 feet (3048 mm) from lot lines or buildings on the same lot.
2. Mechanical and gravity outdoor air intake openings shall be located not less than 10 feet (3048 mm) horizontally from any hazardous or noxious contaminant source, such as vents, streets, alleys, parking lots and loading docks, except as specified in Item 3 or Section 501.3.1. Outdoor air intake openings shall be permitted to be located less than 10 feet (3048 mm) horizontally from streets, alleys, parking lots and loading docks provided that the openings are located not less than 25 feet (7620 mm) vertically above such locations. Where openings front on a street or public way, the distance shall be measured from the closest edge of the street or public way.
3. Intake openings shall be located not less than 3 feet (914 mm) below contaminant sources where such sources are located within 10 feet (3048 mm) of the opening. Separation is not required between intake air openings and living space exhaust air openings of an individual dwelling unit or sleeping unit where an approved factory-built intake/exhaust combination termination fitting is used to separate the air streams in accordance with the manufacturer's instructions.
4. [No change to text]

501.3.1 Location of exhaust outlets. The termination point of exhaust outlets and ducts discharging to the outdoors shall be located with the following minimum distances:

1. For ducts conveying explosive or flammable vapors, fumes or dusts: 30 feet (9144 mm) from property lines; 10 feet (3048 mm) from operable openings into buildings; 6 feet (1829 mm) from exterior walls and roofs; 30 feet (9144 mm) from combustible walls and operable openings into buildings that are in the direction of the exhaust discharge; 10 feet (3048 mm) above adjoining grade.
2. For other product-conveying outlets: 10 feet (3048 mm) from the property lines; 3 feet (914 mm) from exterior walls and roofs; 10 feet (3048 mm) from operable openings into buildings; 10 feet (3048 mm) above adjoining grade.
3. For all environmental air exhaust: 3 feet (914 mm) from property lines; 3 feet (914 mm) from operable openings into buildings for all occupancies other than Group U; and 10 feet (3048 mm) from mechanical air intakes. Such exhaust shall not be considered hazardous or noxious. Separation is not required between intake air

openings and living space exhaust air openings of an individual dwelling unit or sleeping unit where an approved factory-built intake/exhaust combination termination fitting is used to separate the air streams in accordance with the manufacturer's instructions.

4. Exhaust outlets serving structures in flood hazard areas shall be installed at or above the elevation required by Section 1612 of the *International Building Code* for utilities and attendant equipment.
5. For specific systems, see the following sections:
   - **5.1.** Clothes dryer exhaust, Section 504.4.
   - **5.2.** Kitchen hoods and other kitchen exhaust equipment, Sections 506.3.13, 506.4 and 506.5.
   - **5.3.** Dust, stock and refuse conveying systems, Section 511.2.
   - **5.4.** Subslab soil exhaust systems, Section 512.4.
   - **5.5.** Smoke control systems, Section 513.10.3.
   - **5.6.** Refrigerant discharge, Section 1105.7.
   - **5.7.** Machinery room discharge, Section 1105.6.1.

CHANGE SIGNIFICANCE: It is often difficult to maintain the code-prescribed clearances between exhaust openings and air intake openings for reasons such as window placements, proximity to neighboring buildings and dwelling units and lack of sufficient exterior wall area for locating intakes and outlets. The recognition of special factory-built and engineered dual-purpose fittings can solve many clearance problems in new construction and in existing construction where exhaust fans, air intake openings or energy recovery ventilator equipment is being installed. Such fittings are not intended to be "home-made" shop- or field-constructed fittings, rather they are intended to be engineered designs that are factory-built. The code official has approval discretion for such fittings. Such fittings are designed to severely limit the amount of exhaust air that can be entrained in the intake airflow. Also, combination termination fittings allow a single exterior wall penetration as opposed to two separate exhaust and intake penetrations. There are benefits to reducing wall penetrations including less labor, less envelope air infiltration, less potential for moisture penetration and improved aesthetics.

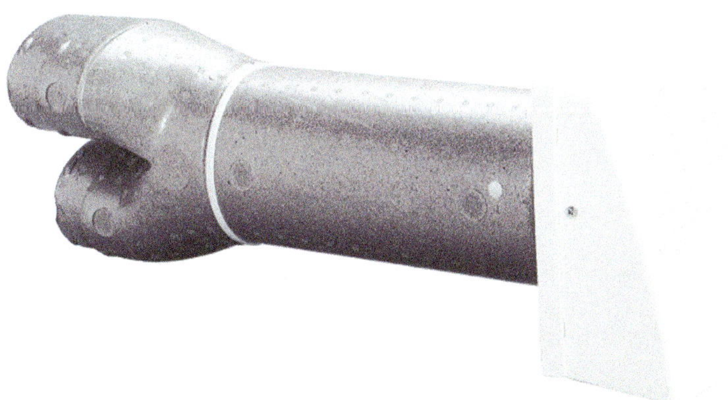

Photo courtesy of Broan-NuTone LLC

Combination exhaust and intake air fitting

Table 403.3.1.1

Recirculation of Mechanical Exhaust Prohibited

CHANGE TYPE: Clarification

CHANGE SUMMARY: Note g of Table 403.3.1.1 was rewritten to lessen the negative impact of recirculated exhaust air from spaces such as bath and toilet rooms, shower rooms, locker rooms and certain classrooms, shops and labs.

2021 CODE:

Table 403.3.1.1
Minimum Ventilation Rates
[No changes to table addressed]

[No changes to notes a through f]

g. Mechanical exhaust is required and recirculation from such spaces is ~~prohibited except that recirculation shall be permitted where the resulting supply airstream consists of not more than 10 percent air recirculated from these spaces.~~ prohibited. For occupancies other than science laboratories, where there is a wheel-type energy recovery ventilation (ERV) unit in the exhaust system design, the volume of air leaked from the exhaust airstream into the outdoor airstream within the ERV shall be less than 10 percent of the outdoor air volume. Recirculation of air that is contained completely within such spaces shall not be prohibited (see Section 403.2.1, Items 2 and 4).

[No change to note h]

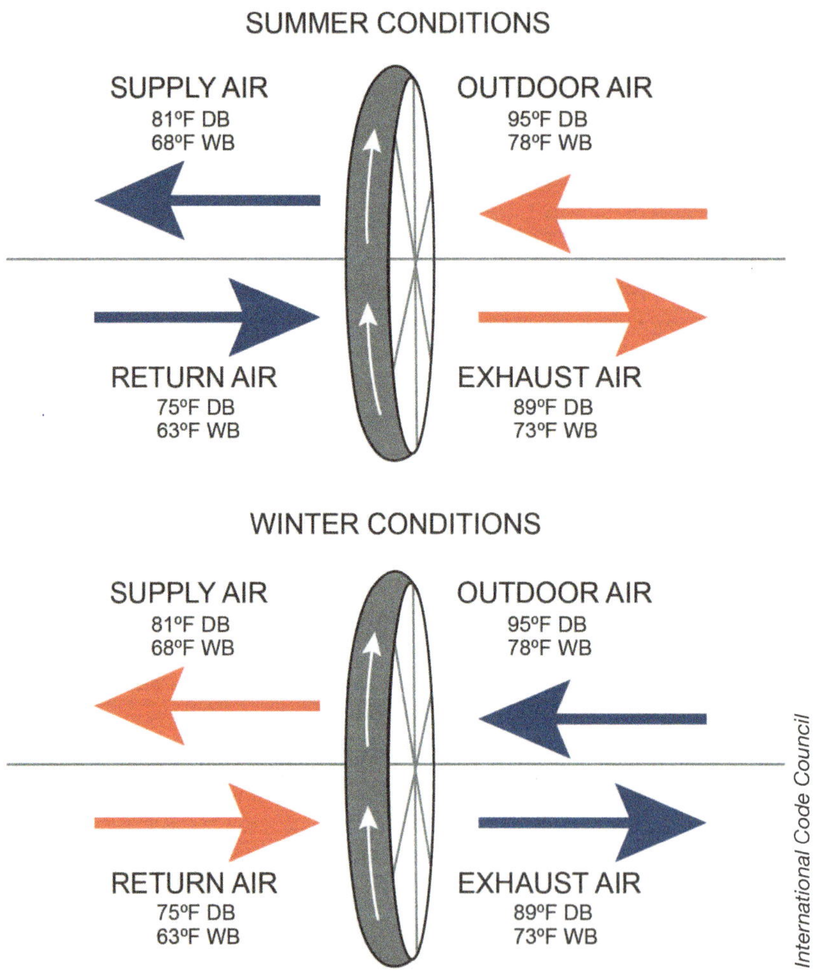

Wheel-type energy recovery ventilation (ERV)

CHANGE SIGNIFICANCE: The previous text of note g would have allowed 10 percent of the entire HVAC supply airflow to any space to be air that was taken from rooms designated by note g in Table 403.3.1.1. The intent of note g was to prevent air from those designated spaces from being circulated to other spaces, except for the small volume of air that leaks across the heat exchanger (cross-leakage) in an energy recovery ventilation (ERV) unit. The revised text clarifies the original intent, which was to allow only a small amount of air to leak from the exhaust airstream of a wheel type ERV into the outdoor airstream of that ERV. The intent of this section is now consistent with ASHRAE 62.1, which is the basis for Table 403.3.1.1.

Wheel type ERVs have a circular heat exchanger that rotates between two airstreams and some cross-leakage and carry-over from one airstream to the other is unavoidable. Sensible heat energy and moisture can be exchanged in an ERV in either direction between an exhaust air duct and an outdoor makeup air duct. Typically, ERVs handle only the outdoor air portion of the total supply air to a space.

Note that the new text intends to prohibit all cross-leakage in ERVs that serve science laboratories if the supply air is to be circulated to some other space. The phrase that begins the second sentence, "For occupancies other than science laboratories," means that the acceptability of limited cross-leakage for ERVs is not acceptable for ERVs that serve science labs. Although the text could be misread to mean that the limit on ERV cross-leakage does not apply to science lab occupancies, the actual intent is that zero cross-leakage is allowed and no amount of air from a science lab is allowed to be circulated to other spaces. This, of course, means that science labs could not be served by wheel-type ERVs if the supply air from the ERV is circulated to any space other than a science lab. Note the last sentence of Note g that clarifies that there is no restriction on recirculation of air from an air handler or an ERV if none of that air leaves the space served by the air handler or ERV.

403.3.1.3

Demand Controlled Ventilation

CHANGE TYPE: Modification

CHANGE SUMMARY: The change clarifies that demand control ventilation schemes cannot eliminate all ventilation in a space while that space is expected to be occupied.

2021 CODE: **403.3.1.3 System operation.** The minimum flow rate of outdoor air that the ventilation system must be capable of supplying during its operation shall be permitted to be based on the rate per person indicated in Table 403.3.1.1 and the actual number of occupants present. Where demand-controlled ventilation is employed to adjust the outdoor airflow rate based on the actual number of occupants present, the minimum quantity of outdoor air shall not fall below that determined from the area outdoor airflow rate column of Table 403.3.1.1 during periods when the building is expected to be occupied.

CHANGE SIGNIFICANCE: Section 403.3.1.3 allows for demand control ventilation, which means that the ventilation rate can be adjusted up or down as the number of occupants in a space increases or decreases. The section might be misinterpreted that the ventilation rate could reduce to zero if the number of occupants was zero. However, it was never intended that the outdoor airflow rate per square foot (R_a) prescribed by Table 403.3.1.1 could be overlooked when the occupant load in a space is zero. When the occupant load is zero, the minimum required ventilation rate is determined by the rate per square foot of floor area (R_a) in the next to last column of Table 403.3.1.1 times the net occupiable floor area of the space. Assuming the table has a prescribed airflow rate based on floor area (R_a) for the space in question in the table, there will always be some ventilation being provided, even when the space has no occupants.

Demand controlled ventilation example

1000 ft^2 high school science classroom/laboratory

Per Table 403.3.1.1, the rate per person (R_p) is 10 cfm;

The rate per square foot (R_a) is 0.18 cfm.

Based on the occupant density from the table, 25

Per 1000 ft^2, the total outdoor air flow rate for occupants would be 250 cfm.

The total outdoor air flow rate based on area would be 180 cfm.

When the science lab has fewer occupants or is unoccupied, the demand controlled ventilation of 250 cfm based on persons could be reduced, perhaps to zero; however, the 180 cfm rate based on area must not be reduced during periods when the building is expected to be occupied. during period when the building is closed to occupants, the minimum ventilation required by this new provision is not applicable.

403.3.2.1

Outdoor Air for Dwelling Units

CHANGE TYPE: Modification

CHANGE SUMMARY: Because of the superior performance of balanced ventilation systems (see new definition), the code will grant a reduction in the ventilation rate, recognizing the higher efficiency of balanced ventilation systems.

2021 CODE: **403.3.2.1 Outdoor air for dwelling units.** An outdoor air ventilation system consisting of a mechanical exhaust system, supply system or combination thereof shall be installed for each dwelling unit. Local exhaust or supply systems, including outdoor air ducts connected to the return side of an air handler, are permitted to serve as such a system. The outdoor air ventilation system shall be designed to provide the required rate of outdoor air continuously during the period that the building is occupied. The minimum continuous outdoor airflow rate shall be determined in accordance with Equation 4-9.

$$Q_{OA} = 0.01A_{floor} + 7.5(N_{br} + 1) \qquad \text{(Equation 4-9)}$$

where:

Q_{OA} = outdoor airflow rate, cfm

A_{floor} = floor area, ft^2

N_{br} = number of bedrooms; not to be less than one

Example calculation of ventilation rate and 30% reduction credit

Example: 1500 sq. foot, 3 bedroom home

$Q = 0.01A_{\text{floor}} = 705(N_{br} + 1).$

Outdoor air cfm $-$ (0.01 × 1500) $-$ [7.5 × (3 + 1)]

O.A. cfm = 15 + 30 = 45

Balanced ventilation system means:

45 cfm O.A. supply with 45 cfm exhaust or

45 cfm O.A. supply with 40.5 to 49.5 cfm exhaust

(see definitions of "balanced ventilation")

New exception 2 allows a 30% reduction for systems:

1. That are balanced and
2. That supply outdoor air through ducts to each bedroom and at least one of the following rooms: living room, dining room, kitchen

Balanced ventilation system O.A. cfm = 45

45 × 0.3 = 13.5 cfm

45 − 13.5 = 31.5 cfm

Or

45 × 0.7 = 31.5 cfm

Exceptions:

1. The outdoor air ventilation system is not required to operate continuously where the system has controls that enable operation for not less than 1 hour of each 4-hour period. The average outdoor airflow rate over the 4-hour period shall be not less than that prescribed by Equation 4-9.
2. The minimum mechanical ventilation rate determined in accordance with Equation 4-9 shall be reduced by 30 percent provided that both of the following apply:
   - **2.1.** A ducted system supplies ventilation air directly to each bedroom and to one or more of the following rooms:
     - **2.1.1.** Living room
     - **2.1.2.** Dining room
     - **2.1.3.** Kitchen
   - **2.2.** The whole-house ventilation system is a balanced system.

SECTION 202
GENERAL DEFINITIONS

BALANCED VENTILATION. Any combination of concurrently operating mechanical exhaust and mechanical supply whereby the total mechanical exhaust airflow rate is within 10 percent of the total mechanical supply airflow rate.

CHANGE SIGNIFICANCE: The code will now give credit to dwelling unit ventilation systems that provide better distribution of outdoor air and provide for the mixing of indoor and outdoor air for better dilution of contaminants. The ventilation rate determined by equation 4-9 can be reduced by 30 percent where the ventilation system is balanced and outdoor air is ducted to each bedroom and also to the choice of the living room, dining room or kitchen. Balanced ventilation systems consist of both exhaust and supply systems that simultaneously operate to ensure the required through-flow of ventilation air in the dwelling. As defined, balanced systems have exhaust and supply rates that are nearly equal (i.e. the exhaust flow rate must be not more than 10 percent greater or less than the supply airflow rate) For example, if the supply airflow rate is 100 cfm, the exhaust airflow rate must be a maximum of 110 cfm and a minimum of 90 cfm. The intent is to ensure the required through-flow in the dwelling and to prevent positive and negative pressures with respect to the outdoors. Balanced systems can be more complex than supply-only and exhaust-only systems, but their performance is superior, hence, the 30 percent reduction credit. Commonly, energy recovery ventilators (ERVs) are utilized to supply balanced ventilation, as ERVs simultaneously supply and exhaust air and also employ a heat exchanger to precondition the incoming supply air, substantially reducing the energy penalty associated with ventilation.

502.20

Manicure and Pedicure Station Exhaust System

CHANGE TYPE: Modification

CHANGE SUMMARY: The code now requires the continuous operation of nail salon exhaust systems during business hours.

2021 CODE: 502.20 Manicure and pedicure stations. Manicure and pedicure stations shall be provided with an exhaust system in accordance with Table 403.3.1.1, Note h. Manicure tables and pedicure stations not provided with factory-installed exhaust inlets shall be provided with exhaust inlets located not more than 12 inches (305 mm) horizontally and vertically from the point of chemical application.

502.20.1 Operation. The exhaust system for manicure and pedicure stations shall have controls that operate the system continuously when the space is occupied.

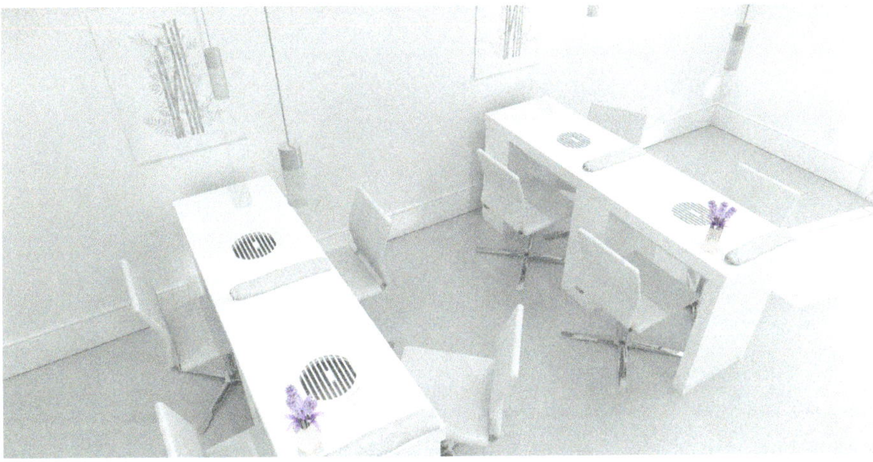

Photo courtesy of SalonSafe LLC

Manicure and pedicure station exhaust system

Photo courtesy of SalonSafe LLC

Manicure and pedicure stations

CHANGE SIGNIFICANCE: It has been observed that the required exhaust systems in nail salons are often not operated as intended, for reasons including fan noise, saving energy and neglecting to turn on the system. Unhealthy levels of contaminants can result in the space if the required exhaust system (source capture system) is not operating when customers are being served. The new code section will require the exhaust system to operate continuously while the space is occupied, which presumably means during the hours the salon is open for business.

504.4.1

Termination Location for Dryer Exhaust

CHANGE TYPE: Modification

CHANGE SUMMARY: New text was added to address the possibility of dryer exhaust air being reintroduced into a building interior.

2021 CODE: **504.4.1 Termination location.** Exhaust duct terminations shall be in accordance with the dryer manufacturer's installation instructions. Where the manufacturer's instructions do not specify a termination location, the exhaust duct shall terminate not less than 3 feet (914 mm) in any direction from openings into buildings including openings in ventilated soffits.

CHANGE SIGNIFICANCE: Previously, the code did not address the required separation between building openings and clothes dryer exhaust terminations. The appliance installation instructions may or may not specify such separation distances. This new section defers to the appliance installation instructions, and if the instructions are silent on the matter, a 3-foot minimum separation distance is specified. Building openings include doors, windows, ventilation and makeup air intake louvers and ventilated soffit intake openings.

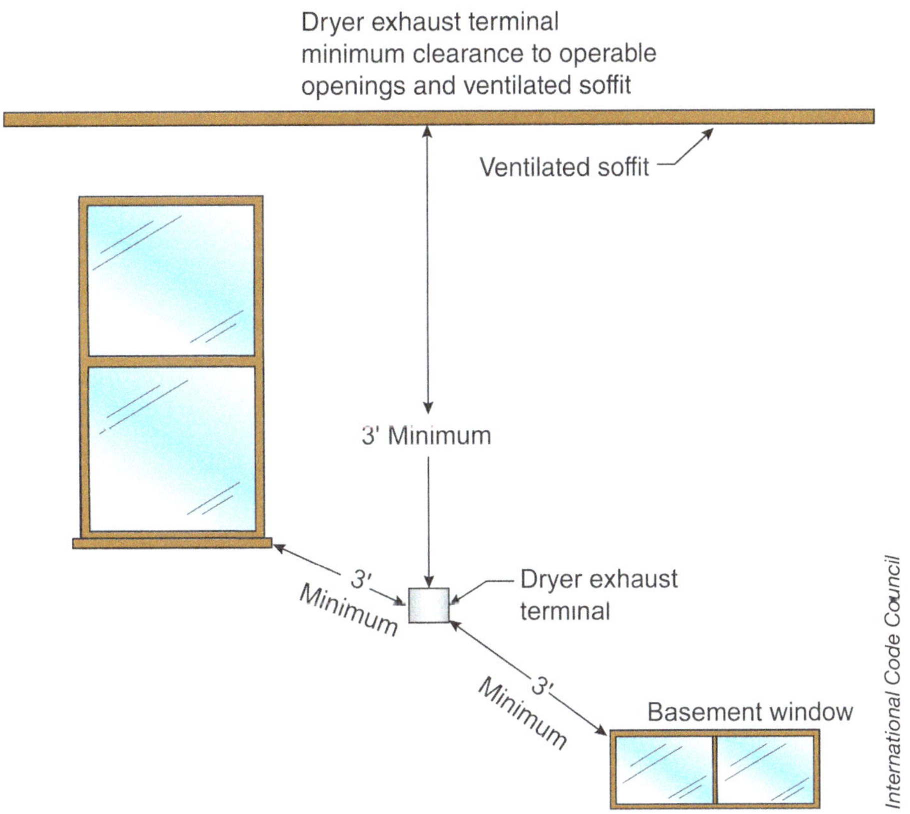

Dryer exhaust terminal minimum clearance to operable openings and soffit vents

506.3.7

Factory-Built Grease Duct Slope

CHANGE TYPE: Modification

CHANGE SUMMARY: The new exception exempts factory-built grease ducts from the duct slope prescriptions of the code, relying instead on the slope requirements stated in the product listing and manufacturer's installation instructions.

2021 CODE: **506.3.7 Prevention of grease accumulation in grease ducts.** Duct systems serving a Type I hood shall be constructed and installed so that grease cannot collect in any portion thereof, and the system shall slope not less than one-fourth unit vertical in 12 units horizontal (2-percent slope) toward the hood or toward a grease reservoir designed and installed in accordance with Section 506.3.7.1. Where horizontal ducts exceed 75 feet (22 860 mm) in length, the slope shall be not less than one unit vertical in 12 units horizontal (8.3-percent slope).

> **Exception:** Factory-built grease ducts shall be installed at a slope that is in accordance with the listing and manufacturer's installation instructions.

International Code Council

Grease duct

CHANGE SIGNIFICANCE: Factory-built grease ducts are tested to UL 1978. and the required minimum slope is determined by testing. Unlike rectangular and square grease ducts, factory-built grease ducts are round and the invert of such horizontal ducts is a curved channel with a much smaller surface area over which grease would flow. It is logical for round ducts to have less slope than flat bottom horizontal ducts. Having less slope will ease installation restraints for long horizontal duct runs.

International Code Council

Grease duct

507.1

Smoker Ovens with Integral Exhaust

CHANGE TYPE: Modification

CHANGE SUMMARY: The new exception exempts smoker ovens from the requirement for a Type I hood where such ovens have an integral exhaust system and are listed for installation without a Type I hood.

2021 CODE: 507.1 General. Commercial kitchen exhaust hoods shall comply with the requirements of this section. Hoods shall be Type I or II and shall be designed to capture and confine cooking vapors and residues. A Type I or Type II hood shall be installed at or above appliances in accordance with Sections 507.2 and 507.3. Where any cooking appliance under a single hood requires a Type I hood, a Type I hood shall be installed. Where a Type II hood is required, a Type I or Type II hood shall be installed. Where a Type I hood is installed, the installation of the entire system, including the hood, ducts, exhaust equipment and makeup air system shall comply with the requirements of Sections 506, 507, 508 and 509.

Photo courtesy of J & R Manufacturing, Inc.

Smoker oven with integral exhaust

Exceptions:

1. Factory-built commercial exhaust hoods that are listed and labeled in accordance with UL 710, and installed in accordance with Section 304.1, shall not be required to comply with Sections 507.1.5, 507.2.3, 507.2.5, 507.2.8, 507.3.1, 507.3.3, 507.4 and 507.5.
2. Factory-built commercial cooking recirculating systems that are listed and labeled in accordance with UL 710B, and installed in accordance with Section 304.1, shall not be required to comply with Sections 507.1.5, 507.2.3, 507.2.5, 507.2.8, 507.3.1, 507.3.3, 507.4 and 507.5. Spaces in which such systems are located shall be considered to be kitchens and shall be ventilated in accordance with Table 403.3.1.1. For the purpose of determining the floor area required to be ventilated, each individual appliance shall be considered as occupying not less than 100 square feet (9.3 m^2).
3. Where cooking appliances are equipped with integral downdraft exhaust systems and such appliances and exhaust systems are listed and labeled for the application in accordance with NFPA 96, a hood shall not be required at or above them.

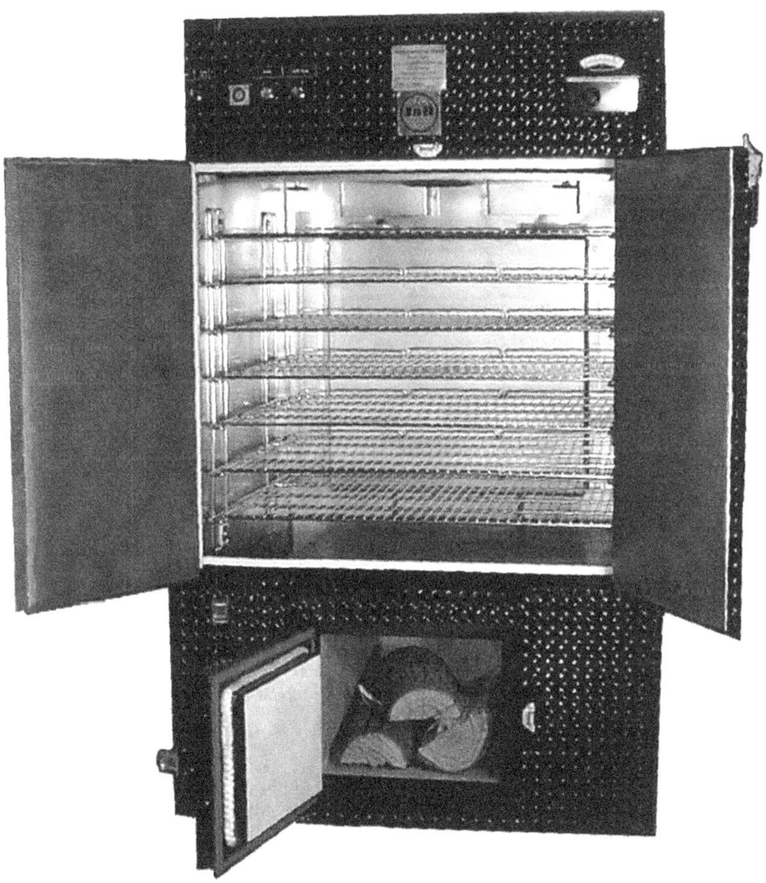

Photo courtesy of J & R Manufacturing, Inc.

Smoker oven with integral exhaust

4. Smoker ovens with integral exhaust systems, provided that the appliance is installed in accordance with the manufacturer's installation instructions, is listed and tested for the application and complies with Chapter 5.

CHANGE SIGNIFICANCE: Cooking appliances that burn solid fuel with an open flame occurring at any time are defined as extra-heavy-duty appliances and Section 507.2 requires a Type I exhaust hood for such appliances. The new exception allows smoker ovens to be exhausted by means of an integral exhaust system that discharges to a factory-built chimney or duct system in accordance with the listing of the appliance. Note that some appliances offer alternatives to a Type I hood, but others specify that a Type I hood must be used over the appliance. This new Exception 4 is specific to smoker ovens that have integral (built-in) exhaust systems that vent the cooking effluent to the outdoors, that assist with draft or that evacuate smoke from the appliance interior before the doors are opened. The appliances that fall under the new exception must be listed and labeled as suitable for venting by means other than a Type I hood.

514.2

Energy Recovery Ventilation Systems

CHANGE TYPE: Modification

CHANGE SUMMARY: The prohibition of some types of energy recovery ventilation (EVR) equipment for use with Type II kitchen exhaust hoods has been lifted.

2021 CODE: **514.2 Prohibited applications.** Energy recovery ventilation systems shall not be used in the following systems:

1. Hazardous exhaust systems covered in Section 510.
2. Dust, stock and refuse systems that convey explosive or flammable vapors, fumes or dust.
3. Smoke control systems covered in Section 513.
4. Commercial kitchen exhaust systems serving Type I ~~or Type II~~ hoods.
5. Clothes dryer exhaust systems covered in Section 504.

Exception: The application of ERV equipment that recovers sensible heat only utilizing coil-type heat exchangers shall not be limited by this section.

CHANGE SIGNIFICANCE: Commercial kitchen exhaust systems are a rich source of energy that could be captured and put to good use rather than wasted to the atmosphere. The concern for the use of ERVs with kitchen exhaust is that the exhaust could foul the ERV heat exchanger and could increase the risk of grease fires. The reference to Type II hoods in Item 4 was deleted; therefore, the sensible and latent heat in the exhaust from Type II hoods serving dishwashers and appliances that do not

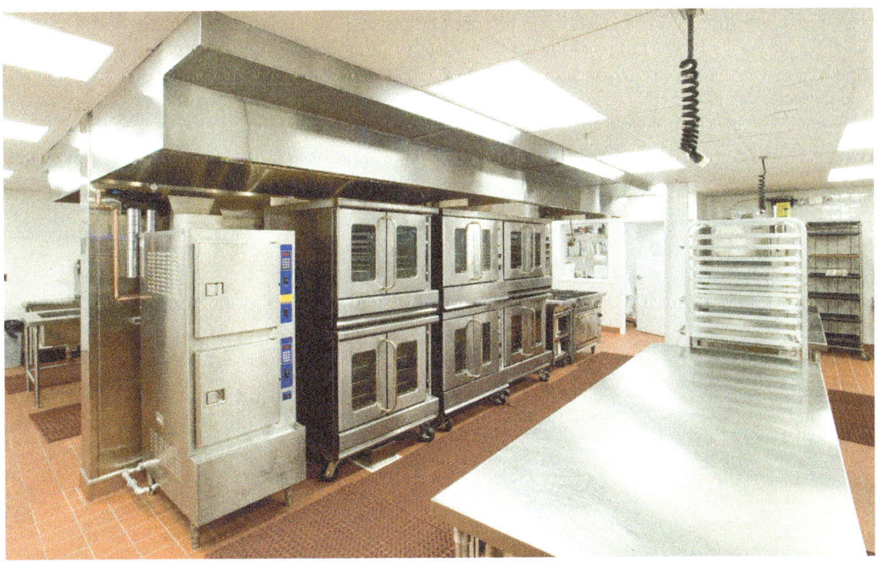

Photo courtesy of gettyimages.com/alacatr

Commercial kitchen ovens

produce grease or smoke can now be recovered by any appropriate type of ERV (i.e. not limited to just simple coil-type sensible heat exchangers). The exhaust effluent from Type II hoods consists of waste heat and water vapor and is not expected to be harmful to ERV equipment. Conversely, the exhaust effluent from Type I hoods typically bears grease, smoke, creosote, particulate matter and other cooking byproducts that would be detrimental to ERV equipment and that could create a fuel load for potential fires.

Photo courtesy of gettyimages.com/onfilm

Commercial kitchen ovens

602.2.1.8

Pipe and Duct Insulation Within Plenums

CHANGE TYPE: Modification

CHANGE SUMMARY: The revision addresses the practice of using pipe insulation materials to protect piping that does not meet the required fire performance requirements.

2021 CODE: **602.2.1.8 Pipe and duct insulation within plenums.** Pipe and duct insulation contained within plenums, including insulation adhesives, shall have a flame spread index of not more than 25 and a smoke developed index of not more than 50 when tested in accordance with ASTM E84 or UL 723, using the specimen preparation and mounting procedures of ASTM E2231. Pipe and duct insulation shall not flame, glow, smolder or smoke when tested in accordance with ASTM C411 at the temperature to which they are exposed in service. The test temperature shall not fall below 250°F (121°C). Pipe and duct insulation shall be listed and labeled. Pipe and duct insulation shall not be used to reduce the maximum flame spread and smoke-developed indices except where the pipe or duct and its related insulation, coatings and adhesives are tested as a composite assembly in accordance with Section 602.2.1.7.

Pipe insulation

CHANGE SIGNIFICANCE: It is common for plastic piping to be installed in plenum spaces, such as above-ceiling return air plenums. Section 602.2.1 requires that materials installed within plenums be noncombustible or have a flame spread index of not more than 25 and a smoke-developed index of not more than 50 when tested in accordance with ASTM E84 or UL 723. Some plastic piping materials cannot meet the requirements of Section 602.2.1; therefore, if they are installed within a plenum, they must be protected by some material that does meet the fire performance requirements and that is listed for that application. Section 602.2.1 allows combustible piping to be "fully enclosed within materials listed and labeled for installation within a plenum and listed for the application." The revised code text states that if pipe insulation is used as the enclosing material, the entire assembly of the pipe, the insulation and any coverings and adhesives must be tested as a composite assembly. Although there are insulation materials available that meet the fire performance requirements themselves, it is an unknown how the entire assembly will perform unless the entire assembly is tested. Without being tested as a composite assembly, it is possible that the protected piping could break down and degrade when exposed to heat, despite being protected by a pipe insulation material.

Photo courtesy of gettyimages.com/Ladsritha

Duct insulation

607.5.2

Duct Penetrations of Fire Barriers

CHANGE TYPE: Modification

CHANGE SUMMARY: The revision describes how flexible air connectors can be installed in a duct system that is otherwise required to be constructed entirely of sheet steel.

2021 CODE: **607.5.2 Fire barriers.** Ducts and air transfer openings that penetrate fire barriers shall be protected with listed fire dampers installed in accordance with their listing. Ducts and air transfer openings shall not penetrate enclosures for interior exit stairways and ramps and exit passageways except as permitted by Sections 1023.5 and 1024.6, respectively, of the *International Building Code.*

> **Exception:** Fire dampers are not required at penetrations of fire barriers where any of the following apply:
>
> 1. Penetrations are tested in accordance with ASTM E119 or UL 263 as part of the fire-resistance-rated assembly.
> 2. Ducts are used as part of an approved smoke control system in accordance with Section 513 and where the fire damper would interfere with the operation of the smoke control system.
> 3. Such walls are penetrated by fully ducted HVAC systems, have a required fire-resistance rating of 1 hour or less, are in areas of other than Group H and are in buildings equipped throughout with an automatic sprinkler system in accordance with Section 903.3.1.1 or 903.3.1.2 of the *International Building Code.* For the purposes of this exception, a fully ducted HVAC system shall be a duct system for the structure's HVAC system. Such a duct system shall be constructed of sheet steel not less than 26 gage [0.0217 inch (0.55 mm)] thickness and shall be continuous from the air-handling appliance or equipment to the air outlet and inlet terminals. Flexible air connectors shall be permitted in a fully ducted system, limited to the following installations.
>   - 3.1. Nonmetallic flexible connections that connect a duct to an air handling unit or equipment located within a mechanical room in accordance with Section 603.9.
>   - 3.2. Nonmetallic flexible air connectors in accordance with Section 603.6.2 that connect an overhead metal duct to a ceiling diffuser where the metal duct and ceiling diffuser are located within the same room.

CHANGE SIGNIFICANCE: Exception 3 allows fire barrier walls to be penetrated by ducts without fire damper protection under the conditions stated in the exception. One of the conditions is that the HVAC system must be a fully ducted system constructed entirely of 26 gage or heavier sheet steel from the air handlers to all outlet and inlet terminals. This condition precludes the use of flexible duct and flexible air connectors as they are not constructed of sheet steel and are typically of nonmetallic construction.

The new text states the conditions under which flexible air connectors (defined in the IMC) can be installed in a fully ducted HVAC system. Flexible air connectors are limited to 14 feet in length and are intended for connecting metal ducts to ceiling air terminals for both supply and return air. Such connectors are practically a necessity because the exact location of ceiling air terminals in a suspended ceiling grid is an unknown at the time when the steel ducts are installed above the ceiling. Item 3.2 of Exception 3 sets three conditions that must be met for the allowance of flexible air connectors in a fully ducted HVAC system.

1. The flexible air connectors are subject to Sections 603.6.2 through 603.6.2.2 which require listing and labeling to UL 181; limit the length to 14 feet and prohibit the connectors from passing through any wall, floor or ceiling.
2. The flexible air connectors are installed only above the ceiling to connect metal ducts to ceiling diffusers.
3. The metal duct, the ceiling diffuser and the flexible air connector are all located in the same room.

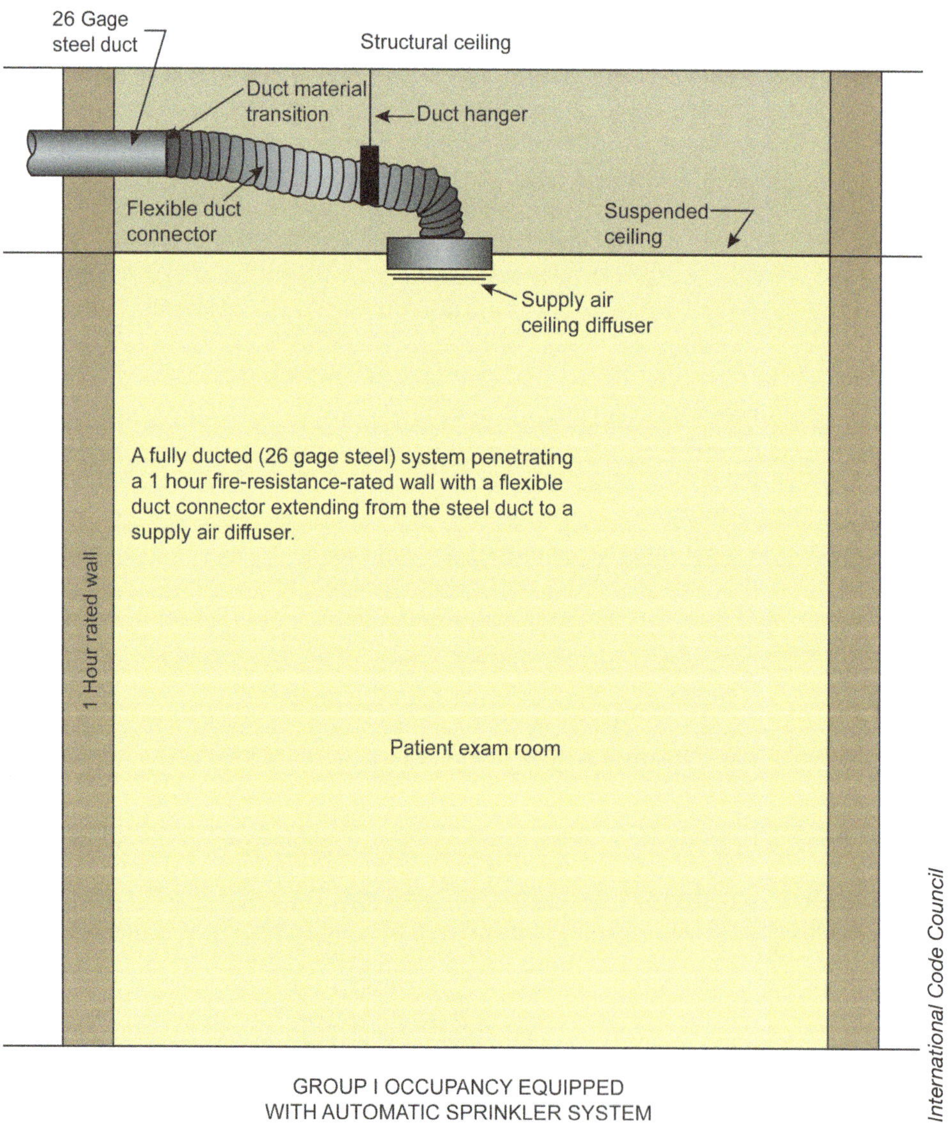

Duct penetration of fire barrier

The logic behind this allowance is that if a metal duct is allowed to serve an intake or supply opening in a room, there is no consequence in having a section of flexible air connector between that metal duct and that opening in the ceiling. With or without the flexible air connector, the metal duct is still connected to an opening in the ceiling; therefore, the presence of a flexible component changes nothing.

For example, consider a 26-gage sheet steel duct that passes through a fire barrier without a fire damper and terminates at a ceiling register three feet inside a room. The intent of Section 607.5.2, Exception 3 has been met because the wall penetration is made by a sheet steel duct. Installing a flexible air connector to allow that same ceiling register to be relocated elsewhere in that same room has no effect on the integrity of the fully ducted HVAC system.

Note that Item 3.1 of Exception 3 does not appear to be related to flexible air connectors, but rather, appears to be referring to the flexible material vibration isolators commonly used to connect ductwork to air handlers. Flexible air connectors were not intended for such applications.

607.5.5

Subducts Penetrating Shaft Enclosures

CHANGE TYPE: Modification

CHANGE SUMMARY: The revision makes Exception 1 consistent with Exception 2 because the requirements for a subduct option for fire dampers should be the same for the subduct option for smoke dampers.

2021 CODE: **607.5.5 Shaft enclosures.** Shaft enclosures that are permitted to be penetrated by ducts and air transfer openings shall be protected with listed fire and smoke dampers installed in accordance with their listing.

Exceptions:

1. Fire dampers are not required at penetrations of shafts where any of the following apply:
   - **1.1.** Steel exhaust subducts having a wall thickness of not less than 0.0187 inch (0.4712 mm) extend not less than 22 inches (559 mm) vertically in exhaust shafts and an exhaust fan is installed at the upper terminus of the shaft that is powered continuously, in accordance with Section 909.11 of the *International Building Code*, so as to maintain a continuous airflow upward to the outdoors. ~~provided that there is a continuous airflow upward to the outdoors.~~
   - **1.2.** Penetrations are tested in accordance with ASTM E119 or UL 263 as part of the fire-resistance-rated assembly.
   - **1.3.** Ducts are used as part of an approved smoke control system in accordance with Section 909 of the *International Building Code*, and where the fire damper will interfere with the operation of the smoke control system.
   - **1.4.** The penetrations are in parking garage exhaust or supply shafts that are separated from other building shafts by not less than 2-hour fire-resistance-rated construction.

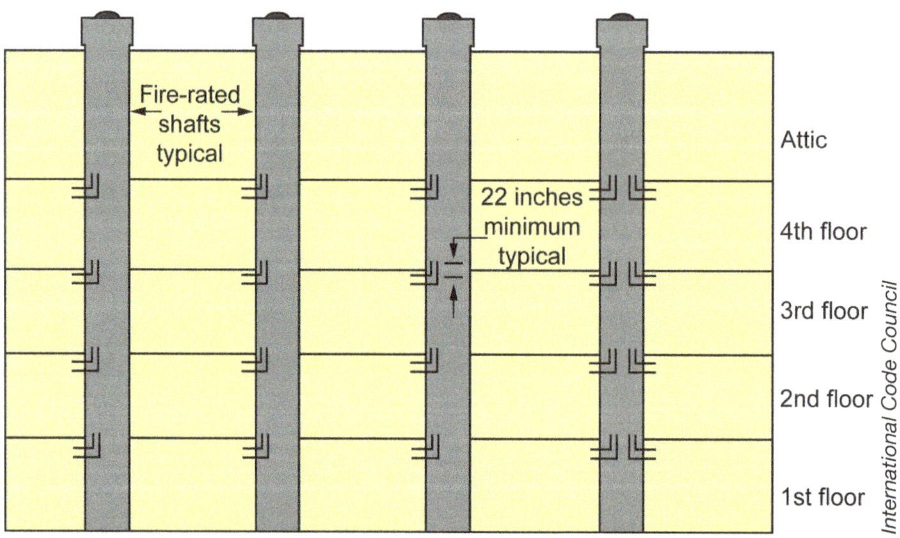

Subducts penetrating shaft enclosures

2. In Group B and R occupancies equipped throughout with an automatic sprinkler system in accordance with Section 903.3.1.1 of the *International Building Code*, smoke dampers are not required at penetrations of shafts where kitchen, clothes dryer, bathroom and toilet room exhaust openings with steel exhaust subducts, having a ~~minimum~~ wall thickness of not less than 0.0187 inch (0.4712 mm) ~~(No. 26 gage),~~ extend not less than 22 inches (559 mm) vertically and the exhaust fan at the upper terminus is powered continuously in accordance with the provisions of Section 909.11 of the *International Building Code*, and maintains airflow upward to the outdoors.
3. Smoke dampers are not required at penetrations of exhaust or supply shafts in parking garages that are separated from other building shafts by not less than 2-hour fire-resistance-rated construction.
4. Smoke dampers are not required at penetrations of shafts where ducts are used as part of an approved mechanical smoke control system designed in accordance with Section 909 of the *International Building Code* and where the smoke damper will interfere with the operation of the smoke control system.
5. Fire dampers and combination fire/smoke dampers are not required in kitchen and clothes dryer exhaust systems where dampers are prohibited by this code.

CHANGE SIGNIFICANCE: Exception 1 provides an option for subducts instead of fire dampers; however, the requirements for the subduct option were incomplete compared to the same option for smoke dampers in Exception 2. Namely, Exception 1 did not state a minimum wall thickness for the subduct steel, it did not refer to a continuously operating exhaust fan at the shaft terminus as the means for creating the upward airflow in the shaft, and it did not require the exhaust fan to be provided with standby power. Exception 1 lacked some of the important requirements for subducts that were in Exception 2, even though the subduct options for fire dampers and smoke dampers both rely on the exact same principles. This revision corrects an oversight in the code.

608.1, 403.3.1.5

Ventilation Air Distribution

CHANGE TYPE: Modification

CHANGE SUMMARY: This revision relocates text that requires verification of the required ventilation airflow rates by means of balancing the system and adds a requirement for airflow adjustment means for air distribution and exhaust systems, in addition to the previous requirement for ventilation air systems.

2021 CODE: ~~**403.3.1.5 Balancing.** The ventilation air distribution system shall be provided with means to adjust the system to achieve not less than the minimum ventilation airflow rate as required by Sections 403.3 and 403.3.1.2. Ventilation systems shall be balanced by an approved method. Such balancing shall verify that the ventilation system is capable of supplying and exhausting the airflow rates required by Sections 403.3 and 403.3.1.2.~~

608.1 Balancing. Air distribution, ventilation and exhaust systems shall be provided with means to adjust the system to achieve the design airflow rates and shall be balanced by an approved method. Ventilation air distribution shall be balanced by an approved method and such balancing shall verify that the air distribution system is capable of supplying and exhausting the airflow rates required by Chapter 4.

CHANGE SIGNIFICANCE: Chapter 6 is the more logical place for balancing requirements. The requirement for an adjustment means should have been broadly applicable to all air conveyance systems, not just ventilation air systems. However, the code was silent on balancing for systems other than ventilation air systems. The new section will now require the balancing of all air conveyance systems as opposed to only ventilation air systems. The requirement for verifying the required ventilation airflow rates of Chapter 4 is relocated to the new balancing section to consolidate all balancing requirements in Chapter 6.

Photo courtesy of gettyimages.com/BanksPhotos

Means to adjust airflow to balance ventilation air distribution

801.21

Blocked Vent Switch for Oil-Fired Appliances

CHANGE TYPE: Addition

CHANGE SUMMARY: A requirement was added for an additional safety device for oil-fired appliances to be consistent with what is required for some gas-fired appliances.

2021 CODE: **801.21 Blocked vent switch.** Oil-fired appliances shall be equipped with a device that will stop burner operation in the event that the venting system is obstructed. Such device shall have a manual reset, and shall be installed in accordance with the manufacturer's instructions.

CHANGE SIGNIFICANCE: Blocked (obstructed) vent switches are not new, but the code has not required them until now. Oil-fired appliances such as furnaces vent to masonry chimneys, factory-built chimneys or Type L vents. These venting systems, especially masonry chimneys, are subject to blockage by items including plant debris, animal nesting materials, bird and raccoon carcasses and deteriorated masonry. The required blocked vent switch detects spillage from the venting system, typically from a draft regulator installed between the appliance and the vent or chimney, and shuts off the appliance. This function can prevent carbon monoxide poisoning in the event of an obstructed vent or chimney. The manual reset feature prevents the appliance from operating without the intentional action by an occupant or an HVAC service person. If the blocked vent switch activates, the cause must be determined and remedied before the appliance is allowed to resume operation.

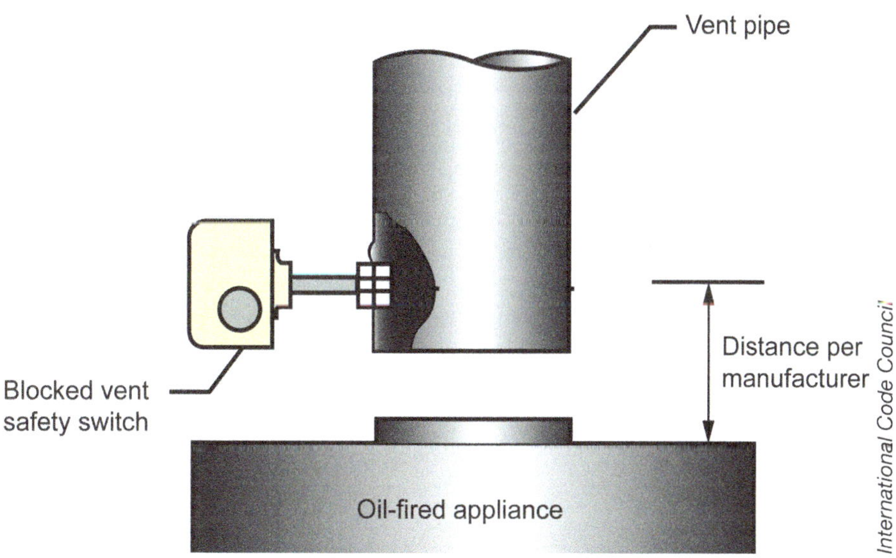

Blocked vent switch for oil-fired appliances

905.1

Wood-Burning Residential Hydronic Heaters

CHANGE TYPE: Modification

CHANGE SUMMARY: The revision makes the code consistent with EPA rules for heater emissions.

2021 CODE: 905.1 General. Fireplace stoves and solid-fuel-type room heaters shall be listed and labeled and shall be installed in accordance with the conditions of the listing. Fireplace stoves shall be tested in accordance with UL 737. Solid-fuel-type room heaters shall be tested in accordance with UL 1482. Fireplace inserts intended for installation in fireplaces shall be listed and labeled in accordance with the requirements of UL 1482 and shall be installed in accordance with the manufacturer's instructions. New wood-burning residential hydronic heaters shall be EPA certified.

CHANGE SIGNIFICANCE: Hydronic heaters are solid fuel-burning hot water boilers. They are popular in rural areas and are also known as outdoor wood furnaces, among other names. They can burn wood, coal, pellets and biomass fuel. Typically, these units are located outdoors, vent through short chimneys attached directly to the unit and operate as unpressurized vessels, at or very near atmospheric pressure. The typical installation involves a hydronic heater installed outdoors on a pad with insulated plastic tubing run underground to a water coil in an air handler located in the building to be heated. These heaters have evolved to become much more efficient and cleaner burning than older technology. Older technology heaters produced much higher particulate emissions, and because the heaters vented through short chimneys, the smoke could stay close to the ground and seriously affect the air quality in the vicinity of such heaters.

The EPA was concerned with improving the air quality in areas where solid fuel was used to heat homes and other buildings and requires that these heaters be tested and certified as complying with the EPA limits on emissions (combustion byproducts). The emission limits will become more stringent in the future. The revised code text requires what the EPA requires to prevent higher polluting appliances from being installed.

The following link addresses the EPA rules for hydronic heaters: https://www.epa.gov/sites/production/files/2015-02/documents/20150204fs-hydronic-heaters.pdf.

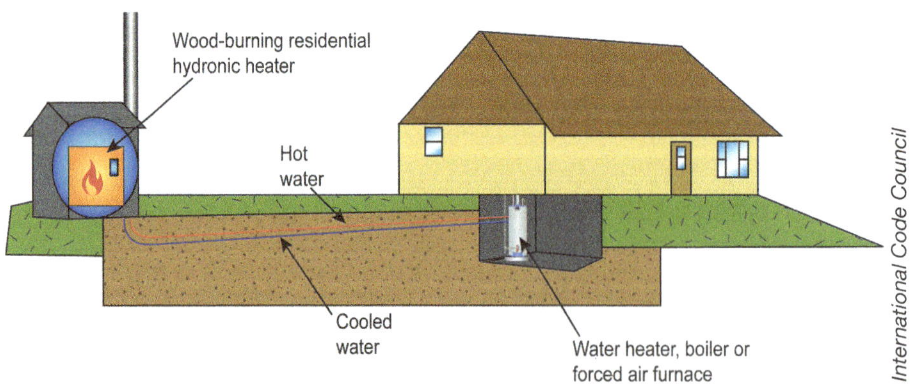

Wood-burning residential hydronic heater

CHANGE TYPE: Addition

CHANGE SUMMARY: Coverage was added for a new type of decorative fuel-burning appliance about which the code was previously silent.

929

Unvented Alcohol Fuel-Burning Decorative Appliances

2021 CODE:

SECTION 929
UNVENTED ALCOHOL FUEL-BURNING DECORATIVE APPLIANCES

929.1 General. Unvented alcohol fuel-burning decorative appliances shall be listed and labeled in accordance with UL 1370 and shall be installed in accordance with the conditions of the listing, manufacturer's instructions and Chapter 3.

SECTION 202
GENERAL DEFINITIONS

UNVENTED ALCOHOL FUEL-BURNING DECORATIVE APPLIANCE. A stationary, self-contained appliance intended to be directly or indirectly secured to a wall or floor and not intended for duct connection. Such appliance burns alcohol and is made in a manufacturing facility for subsequent delivery to the installation site.

CHANGE SIGNIFICANCE: The code did not address unvented alcohol fuel-burning decorative appliances in previous editions, but now they are within the scope of the code. The referenced UL standard 1370 is titled: *Unvented Alcohol Fuel Burning Decorative Appliances*. These appliances are not heaters and perform a strictly decorative function. As stated in the definition, they are stationary (not portable) and must be secured to a wall or floor. The applicable provisions of Chapter 3 apply to these appliances, such as Sections 303 and 304. As with all appliances regulated by the IMC, these decorative appliances must be installed in strict accordance with the manufacturer's instructions and the conditions of the product listing. Installation considerations include, among others, clearances to combustibles, the minimum required room volume and access. The energy input is limited to 0.25 gallon of denatured alcohol fuel per hour.

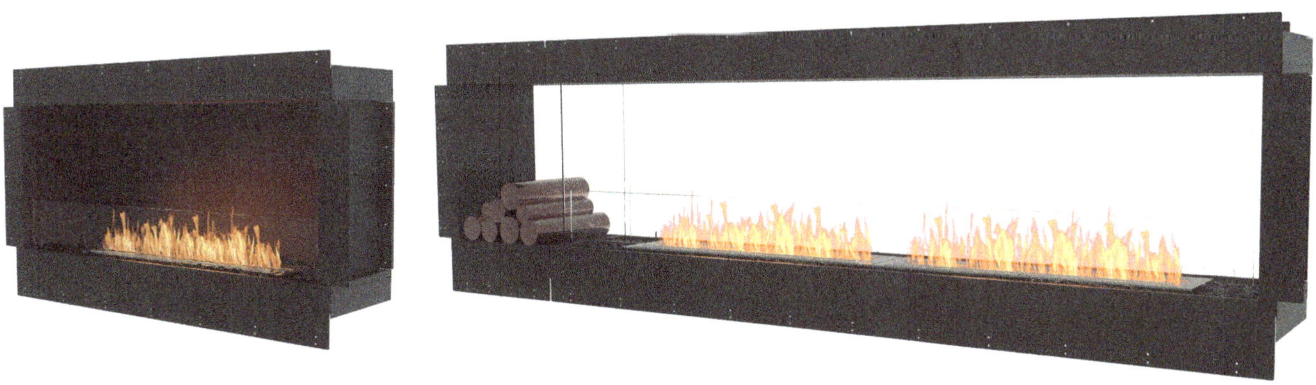

Photo courtesy of EcoSmart Fire

Unvented alcohol fuel-burning decorative appliance

1101 through 1108

Ammonia Refrigeration Systems

CHANGE TYPE: Modification

CHANGE SUMMARY: These revisions remove all regulations for ammonia refrigeration systems from the IMC and instead simply defer all regulation to the suite of IIAR standards already referenced in the IMC.

2021 CODE: **1101.1 Scope.** This chapter shall govern the design, installation, construction and repair of refrigeration systems that vaporize and liquefy a fluid during the refrigerating cycle. ~~Refrigerant piping design and installation, including pressure vessels and pressure relief devices, shall conform to this code.~~ Permanently installed refrigerant storage systems and other components shall be considered as part of the refrigeration system to which they are attached.

1101.1.1 Refrigerants other than ammonia. Refrigerant piping design and installation for systems containing a refrigerant other than ammonia, including pressure vessels and pressure relief devices, shall comply with this chapter and ASHRAE 15.

1101.1.2 Ammonia refrigerant. Refrigeration systems using ammonia as the refrigerant shall comply with IIAR 2, IIAR 3, IIAR 4 and IIAR 5 and shall not be required to comply with this chapter.

~~**1101.6 General.** Refrigeration systems shall comply with the requirements of this code and, except as modified by this code, ASHRAE 15. Ammonia-refrigerating systems shall comply with this code and, except as modified by this code, ASHRAE 15, IIAR 2, IIAR 3, IIAR 4 and IIAR 5.~~

Photo courtesy of gettyimages.com/SlavkoSereda

Ammonia refrigeration systems

TABLE 1103.1 Refrigerant Classification, Amount and OEL (only the applicable rows are shown for brevity)

R-32	CH_2F_2	difluoromethane (methylene fluoride)	A2[f,c]	4.8	36,000	77	1,000	1-4-0
R-143A	CH_3CF_3	1,1,1-trifluoroethane	A2[f,c]	4.5	21,000	70	1,000	2-0-0[b]
R-444A	zeotrope	R-32/152a/1234ze(E) (12.0/5.0/83.0)	A2[f,c]	5.1	21,000	81	850	—
R-444B	zeotrope	R-32/152a/1234ze(E) (41.5/10.0/48.5)	A2[f,c]	4.3	23,000	69	890	—
R-445A	zeotrope	R-744/134a/1234ze(E) (6.0/9.0/85.0)	A2[f,c]	4.2	16,000	67	930	—
R-446A	zeotrope	R-32/1234ze(E)/600 (68.0/29.0/3.0)	A2[f,c]	2.5	16,000	39	960	—
R-447A	zeotrope	R-32/125/1234ze(E) (68.0/3.5/28.5)	A2[f,c]	2.6	16,000	42	900	—
R-451A	zeotrope	R-1234yf/134a (89.8/10.2)	A2[f,c]	5.3	18,000	81	520	—
R-451B	zeotrope	R-1234yf/134a (88.8/11.2)	A2[f,c]	5.3	18,000	81	530	—
~~R-717~~	~~NH_3~~	~~ammonia~~	~~B2[f]~~	~~0.014~~	~~320~~	~~0.22~~	~~25~~	~~3-3-0[c]~~
R-1234yf	$CF_3CF{=}CH_2$	2,3,3,3-tetrafluoro-1 propene	A2[f,c]	4.7	16,000	75	500	—
R-1234ze(E)	$CF_3CH{=}CHF$	trans-1,3,3,3-tetrafluoro-1-propene	A2[f,c]	4.7	16,000	75	800	—

a. Degrees of hazard are for health, fire, and reactivity, respectively, in accordance with NFPA 704.

b. Reduction to 1-0-0 is allowed if analysis satisfactory to the code official shows that the maximum concentration for a rupture or full loss of refrigerant charge would not exceed the IDLH, considering both the refrigerant quantity and room volume.

c. ~~For installations that are entirely outdoors, use 3-1-0.~~ The ASHRAE Standard 34 flammability classification for this refrigerant is 2L, which is a subclass of Class 2.

d. Class I ozone depleting substance; prohibited for new installations.

e. Occupational Exposure Limit based on the OSHA PEL, ACGIH TLV-TWA, the TERA WEEL or consistent value on a time-weighted average (TWA) basis (unless noted C for ceiling) for an 8 hr/d and 40 hr/wk.

~~f. The ASHRAE Standard 34 flammability classification for this refrigerant is 2L, which is a subclass of Class 2.~~

1105.6.3 Ventilation rate. ~~For other than ammonia systems, the~~ Mechanical ventilation systems shall be capable of exhausting the minimum quantity of air both at normal operating and emergency conditions, as required by Sections 1105.6.3.1 and 1105.6.3.2. ~~The minimum required emergency ventilation rate for ammonia shall be 30 air changes per hour in accordance with IIAR2.~~ Multiple fans or multispeed fans shall be allowed to produce the emergency ventilation rate and to obtain a reduced airflow for normal ventilation.

~~**1105.8 Ammonia discharge.** Pressure relief valves for ammonia systems shall discharge in accordance with ASHRAE 15.~~

1105.8 Emergency pressure control system. ~~Permanently installed refrigeration systems containing more than 6.6 pounds (3 kg) of flammable, toxic or highly toxic refrigerant or ammonia~~ Emergency pressure control systems shall be provided ~~with an emergency pressure control system~~ in accordance with Section 605.10 of the *International Fire Code*.

~~**1106.3 Ammonia room ventilation.** Ventilation systems in ammonia machinery rooms shall be operated continuously at the ventilation rate specified in Section 1105.6.3.~~

~~**Exceptions:**~~

~~1. Machinery rooms equipped with a vapor detector that will automatically start the ventilation system at the ventilation rate specified in Section 1105.6.3, and that will actuate an alarm at a detection level not to exceed 1,000 ppm.~~

~~2. Machinery rooms conforming to the Class 1, Division 2, hazardous location classification requirements of NFPA 70.~~

1106.3 Flammable refrigerants. Where refrigerants of Groups A2, A3, B2 and B3 are used, the machinery room shall conform to the Class 1, Division 2, hazardous location classification requirements of NFPA 70.

Exceptions:

~~1. Ammonia machinery rooms that are provided with ventilation in accordance with Section 1106.3.~~

~~2.~~ Machinery rooms for systems containing Group A2L refrigerants that are provided with ventilation in accordance with Section 1106.4.

~~**1108.2 Test gases.** Tests shall be performed with an inert dried gas including, but not limited to, nitrogen and carbon dioxide. Oxygen, air, combustible gases and mixtures containing such gases shall not be used.~~

~~**Exception:** The use of air is allowed to test R-717, ammonia, systems provided that they are subsequently evacuated before charging with refrigerant.~~

[Note: Not all changes to Sections 1101 through 1108 are shown. Please refer to the 2021 IMC for the complete text and associated tables.]

CHANGE SIGNIFICANCE: The IIAR standards already referenced in the IMC comprehensively cover all aspects of refrigeration systems that use ammonia as the refrigerant. Therefore, there is no longer the need to have duplicate/overlapping coverage in the IMC that could create confusion and future correlation issues. The IIAR standards referenced in the IMC are stand-alone documents that do not need to rely on the IMC for coverage of ammonia systems. These revisions to the IMC are intended to parallel revisions to ASHRAE 15.

1105.9

Machinery Room Means of Egress

CHANGE TYPE: Addition

CHANGE SUMMARY: Revised egress requirements for machinery rooms from the IBC were added to the IMC to prevent such requirements from being overlooked.

2021 CODE: **1105.9 Means of egress.** Machinery rooms larger than 1,000 square feet ($93\ m^2$) in area shall have not less than two exits or exit access doorways. Where two exit access doorways are required, one such doorway is permitted to be served by a fixed ladder or an alternating tread device. Exit access doorways shall be separated by a horizontal

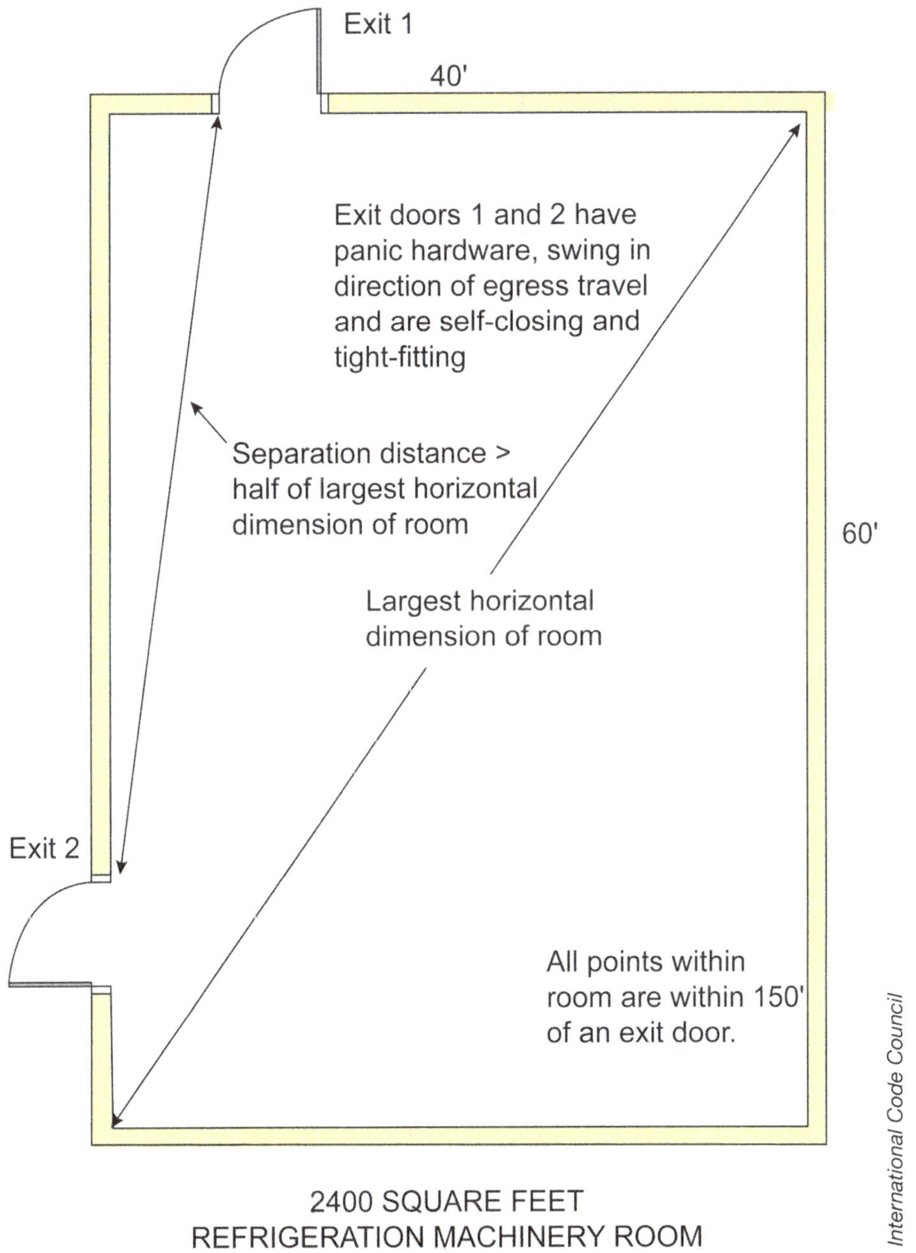

Machinery room means of egress

distance equal to or greater than one-half of the largest horizontal dimension of the room. All portions of machinery rooms shall be within 150 feet (45 720 mm) of an exit or exit access doorway. An increase in exit access travel distance is permitted where in accordance with Section 1017.1 of the *International Building Code*. Exit and exit access doorways shall swing in the direction of egress travel and shall be equipped with panic hardware, regardless of the occupant load served. Exit and exit access doorways shall be tight-fitting and self-closing.

CHANGE SIGNIFICANCE: Egress requirements for machinery rooms are contained in the IBC but were not previously duplicated in the IMC. Machinery room requirements are in the IMC; therefore, egress requirements for such spaces logically belong in the IMC as well. The lack of egress coverage in the IMC could result in the requirements being overlooked by users of the IMC. The IBC egress coverage is added to the IMC and is also revised relative to panic hardware.

It is appropriate for refrigeration machinery rooms to have panic hardware on means of egress doors to protect occupants from the risk of a rapid release of hazardous or asphyxiant gases. The need for rapid escape from refrigeration machinery rooms is not unlike what is needed for Group H occupancies, which are required by the IBC to have panic hardware on all swinging doors. Likewise, IIAR 2 includes this requirement for ammonia refrigeration machinery rooms. The requirement in the IBC is not readily found as a refrigeration machinery room requirement since it is isolated in the means of egress chapter.

Photo courtesy of gettyimages.com/-MG-

Machinery room egress door

1107 through 1110

Refrigerant Piping

CHANGE TYPE: Addition

CHANGE SUMMARY: Section 1107, Refrigerant Piping, of the 2018 IMC was deleted entirely and replaced with all new text in Sections 1107 through 1110.

2021 CODE:

[Due to space constraints, only portions of Sections 1107 through 1110 are shown. Please refer to the 2021 IMC for the complete text and associated tables.]

SECTION 1107
PIPING MATERIAL

1107.1 Piping. Refrigerant piping material for other than R-717 (ammonia) systems shall conform to the requirements in this section. Piping material and installations for R-717 (ammonia) refrigeration systems shall comply with IIAR 2.

1107.2 Used materials. Used pipe, fittings, valves and other materials that are to be reused shall be clean and free of foreign materials and shall be approved for reuse.

1107.3 Material rating. Materials, joints and connections shall be rated for the operating temperature and pressure of the refrigerant system. Materials shall be suitable for the type of refrigerant and type of lubricant in the refrigerant system. Magnesium alloys shall not be used in contact with any halogenated refrigerants. Aluminum, zinc, magnesium and their alloys shall not be used in contact with R-40 (methyl chloride).

International Code Council

Refrigeration piping

TABLE 1107.4 Refrigerant Pipe

Piping Material	Standard (See Chapter 15)

TABLE 1107.5 Refrigerant Pipe Fittings

Fitting Material	Standard (See Chapter 15)

TABLE 1107.5.1 Copper Brazed Swaged Cup Depths

Fitting Size (Inch)	Minimum Depth (Inch)	Maximum Depth (Inch)

SECTION 1108 JOINTS AND CONNECTIONS

1108.1 Approval. Joints and connections shall be of an approved type. Joints and connections shall be tight for the pressure of the refrigerant system when tested in accordance with Section 1110.

1108.1.1 Joints between different piping materials. Joints between different piping materials shall be made with approved adapter fittings. Joints between dissimilar metallic piping materials shall be made with a dielectric fitting or a dielectric union conforming to dielectric tests of ASSE 1079. Adapter fittings with threaded ends between different materials shall be joined with thread lubricant in accordance with Section 1108.3.4.

1108.2 Preparation of pipe ends. Pipe shall be cut square, reamed and chamfered, and shall be free of burrs and obstructions. Pipe ends shall have full-bore openings and shall not be undercut.

1108.3 Joint preparation and installation. Where required by Sections 1108.4 through 1108.9, the preparation and installation of brazed, flared, mechanical, press-connect, soldered, threaded and welded joints shall comply with Sections 1108.3.1 through 1108.3.5.

SECTION 1109 REFRIGERANT PIPE INSTALLATION

1109.1 General. Refrigerant piping installations, other than R-717 (ammonia) refrigeration systems, shall comply with the requirements of this section. The design of refrigerant piping shall be in accordance with ASME B31.5.

1109.2 Piping location. Refrigerant piping shall comply with the installation location requirements of Sections 1109.2.1 through 1109.2.7. Refrigerant piping for Groups A2L and B2L shall also comply with the requirements of Section 1109.3. Refrigerant piping for group A2, A3, B2 and B3 shall also comply with the requirements of Section 1109.4.

1109.2.1 Minimum height. Exposed refrigerant piping installed in open spaces that afford passage shall be not less than 7 feet 3 inches (2210 mm) above the finished floor.

1109.2.2 Refrigerant pipe enclosure. Refrigerant piping shall be protected by locating it within the building elements or within protective enclosures.

> **Exception:** Piping protection within the building elements or protective enclosure shall not be required in any of the following locations:
>
> 1. Where installed without ready access or located more than 7 feet 3 inches (2210 mm) above the finished floor.
> 2. Where located within 6 feet (1829 mm) of the refrigerant unit or appliance.
> 3. Where located in a machinery room complying with Section 1105.

1109.2.3 Prohibited locations. Refrigerant piping shall not be installed in any of the following locations:

1. Exposed within a fire-resistance-rated exit access corridor.
2. Within an interior exit stairway.
3. Within an interior exit ramp.
4. Within an exit passageway.
5. Within an elevator, dumbwaiter or other shaft containing a moving object.

TABLE 1109.3.2 Shaft Ventilation Velocity

Cross-Sectional Area of Shaft (sq. in.)	Minimum Ventilation Velocity (feet per minute)

SECTION 1110
REFRIGERANT PIPING SYSTEM TEST

1110.1 General. Refrigerant piping systems, other than R-717 (ammonia) refrigeration systems, that are erected in the field, shall be pressure tested for strength and leak tested for tightness, in accordance with the requirements of this section, after installation and before being placed in operation. Tests shall include both the high- and low-pressure sides of each system.

Exception: Listed and labeled equipment, including compressors, condensers, vessels, evaporators, gas bulk storage tanks, safety devices, pressure gauges and control mechanisms, shall not be required to be tested.

1110.2 Exposure of refrigerant piping system. Refrigerant pipe and joints installed in the field shall be exposed for visual inspection and testing prior to being covered or enclosed.

1110.3 Test gases. The medium used for pressure testing the refrigerant system shall be one of the following inert gases: oxygen-free nitrogen, helium or argon. For R-744 refrigerant systems, carbon dioxide shall be allowed as the test medium. For R-718 refrigerant systems, water shall be allowed as the test medium. Oxygen, air, combustible gases and mixtures containing such gases shall not be used as test medium. Systems erected on the premises with tubing not exceeding ⅝ inch (15.9 mm) outside diameter shall be allowed to use the refrigerant identified on the nameplate label or marking as the test medium.

1110.4 Test apparatus. The means used to pressurize the refrigerant piping system shall have on its outlet side a test pressure measuring device and either a pressure-limiting device or a pressure-reducing device. The test pressure measuring device shall have an accuracy of ±3 percent or less of the test pressure and shall have a resolution of 5 percent or less of the test pressure.

[Due to space constraints, only portions of Sections 1107 through 1110 are shown. Please refer to the 2021 IMC for the complete text and associated tables.]

Photo courtesy of gettyimages.com/kasipat

Copper refrigeration piping

CHANGE SIGNIFICANCE: The rewrite of Section 1107 reorganizes existing text and updates provisions that have not been changed in many years. The rewrite is more comprehensive than previous code coverage and also includes expanded coverage of some topics and includes several new topics that were not previously covered at all. The new Sections 1107 through 1110 are the culmination of efforts by several experts in the field of refrigerant piping design and installation. The new sections address piping for all refrigerants except ammonia.

SIGNIFICANT CHANGES TO THE

INTERNATIONAL FUEL GAS CODE®

2021 EDITION

PART 3

International Fuel Gas Code

Chapters 1 through 8

- **Chapter 1** Scope and Administration
  No changes addressed
- **Chapter 2** Definitions
- **Chapter 3** General Regulations
- **Chapter 4** Gas Piping Installations
- **Chapter 5** Chimneys and Vents
- **Chapter 6** Specific Appliances
- **Chapter 7** Gaseous Hydrogen Systems
  No changes addressed
- **Chapter 8** Referenced Standards
  No changes addressed

The *International Fuel Gas Code*® (IFGC®) applies to the installation of fuel gas piping systems, fuel gas utilization equipment, gaseous hydrogen systems and related accessories. Chapter 1 provides for the administration and enforcement of the code, assigning responsibility and authority to the code official. Chapter 2 contains definitions of terms specific to their use throughout the code. The general requirement provisions of Chapter 3 govern the approval and installation of all equipment and appliances regulated by the code. Requirements for the design and installation of gas piping systems are set out in Chapter 4 and include provisions for materials, components, fabrication, testing, inspection, operation and maintenance of such systems. The scope of Chapter 5 includes factory-built chimneys, liners, vents, connectors and masonry chimneys serving gas-fired appliances. Reference is made to the *International Mechanical Code*® (IMC®) for chimneys serving appliances using other fuels and to the *International Building Code*® (IBC®) for the construction requirements of masonry chimneys. Approval, design and installation of specific appliances such as furnaces, boilers, water heaters, fireplaces, decorative appliances, room heaters and clothes dryers are covered in Chapter 6. Chapter 7 covers the developing technology of gaseous hydrogen systems, including hydrogen generation and refueling operations, and provides reference to the applicable provisions of the *International Fire Code*® (IFC®). Chapter 8 provides a complete list of standards referenced in various sections of the code. ■

202

Definition of Point of Delivery

307.2

Concealed Condensate Piping

402.7, 202

Press-Connect Joint

403.8.3

Threaded Joint Sealing

404.5

Fittings in Concealed Locations

503.5.6.1

Chimney Lining

503.8

Venting System Terminal Clearances

503.10.7

Vent Connector Junctions

614.7

Clothes Dryer Makeup Air

618.6

Return Air from Mechanical Room

623.2

Commercial Cooking Appliances

202

Definition of Point of Delivery

CHANGE TYPE: Modification

CHANGE SUMMARY: New definitions clarify terms that are used in the definition of "point of delivery" and a change to the definition of "service shutoff" coordinates with all of the relative definitions.

2021 CODE: **202 POINT OF DELIVERY.** For natural gas systems, the point of delivery is the outlet of the service meter assembly or the outlet of the service regulator or service shutoff valve where a meter is not provided. Where a system shutoff valve is provided ~~at~~ after the outlet of the service meter assembly, such valve shall be considered to be downstream of the point of delivery. For undiluted liquefied petroleum gas systems, the point of delivery shall be considered to be the outlet of the service pressure regulator, exclusive of line gas regulators, in the system.

SERVICE METER ASSEMBLY. The meter, valve, regulator, piping, fittings and equipment installed by the service gas supplier before the point of delivery.

SYSTEM SHUTOFF. A valve installed after the point of delivery to shut off the entire piping system.

VALVE. A device used in piping to control the gas supply to any section of a system of piping or to an appliance.

Service shutoff. A valve, installed by the serving gas supplier between the ~~service meter or~~ source of supply and the point of delivery ~~customer piping system~~, to shut off the entire piping system.

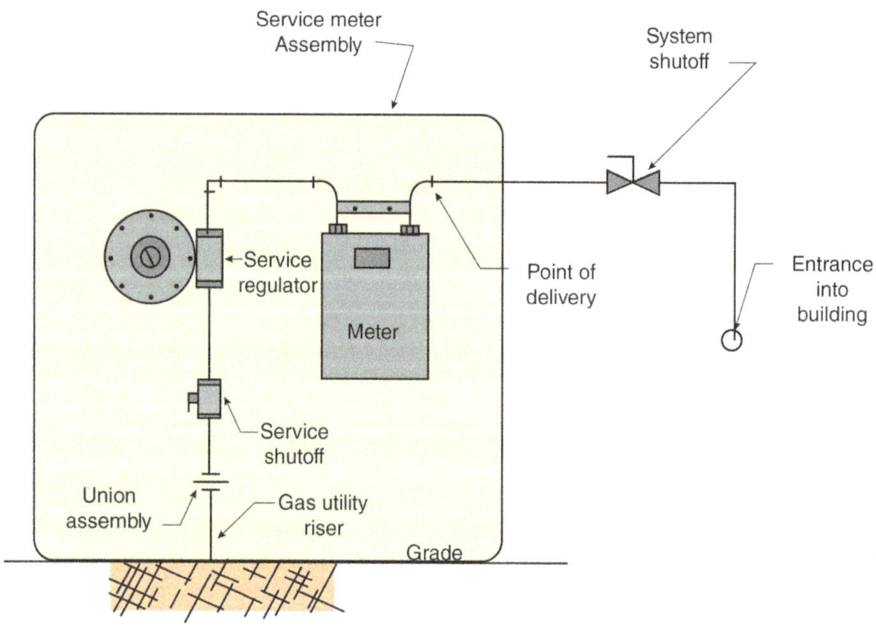

Service meter assembly and point of delivery

International Code Council

CHANGE SIGNIFICANCE: The point of delivery (POD) is the point where the IFGC code applicability begins. Everything downstream of the POD is governed by the IFGC, and everything upstream of the POD is governed by the Department of Transportation (DOT) and the gas utility company. The definition of POD refers to the service meter assembly, but this term was not defined until now. The new definition clarifies which components are under the control of the gas utility company and are, therefore, before (upstream of) the POD and not controlled by the IFGC. The definition of service shutoff (valve) was revised to refer to the POD instead of the vague reference to customer piping system and the term service meter was deleted, recognizing that the new definition of service meter assembly includes the meter as one of several components that are upstream of the POD.

The definition of POD was also revised to include the newly defined term "system shutoff," which is an optional valve installed after (downstream of) the service meter assembly. A system shutoff valve is a valve that is installed by choice to allow the customer or building owner to shut off the gas to the entire building. This valve also allows the gas to be turned on by a party other than the gas utility after the gas utility has set the service meter assembly, thus avoiding the operation of the gas utility owned service shutoff valve. The distinction between a service shutoff valve and a system shutoff valve is that the system shutoff valve is owned and operated by the customer and the service shutoff valve is owned by the gas utility and can be operated only by qualified personnel working for the gas utility. The service shutoff valve is commonly operated by contractors performing work on the building's gas piping system, but, by federal law, this practice is not legal. The service shutoff valve is for use only by gas utility personnel or perhaps emergency personnel. Having a system shutoff valve downstream of the POD will allow contractors to shut off the gas to the building within the law.

307.2

Concealed Condensate Piping

CHANGE TYPE: Modification

CHANGE SUMMARY: The code was revised to address the concern over the terminations of concealed condensate drain lines where it will not be apparent which drain is the normal condensate drain and which drain is the secondary (backup) drain.

2021 CODE: **307.2 Fuel-burning appliances.** Liquid combustion byproducts of condensing appliances shall be collected and discharged to an approved plumbing fixture or disposal area in accordance with the manufacturer's instructions. Condensate piping shall be of approved corrosion-resistant material and shall be not smaller than the drain connection on the appliance. Such piping shall maintain a minimum slope in the direction of discharge of not less than ⅛ unit vertical in 12 units horizontal (1-percent slope). <u>The termination of concealed condensate piping shall be marked to indicate whether the piping is connected to the primary drain or to the secondary drain.</u>

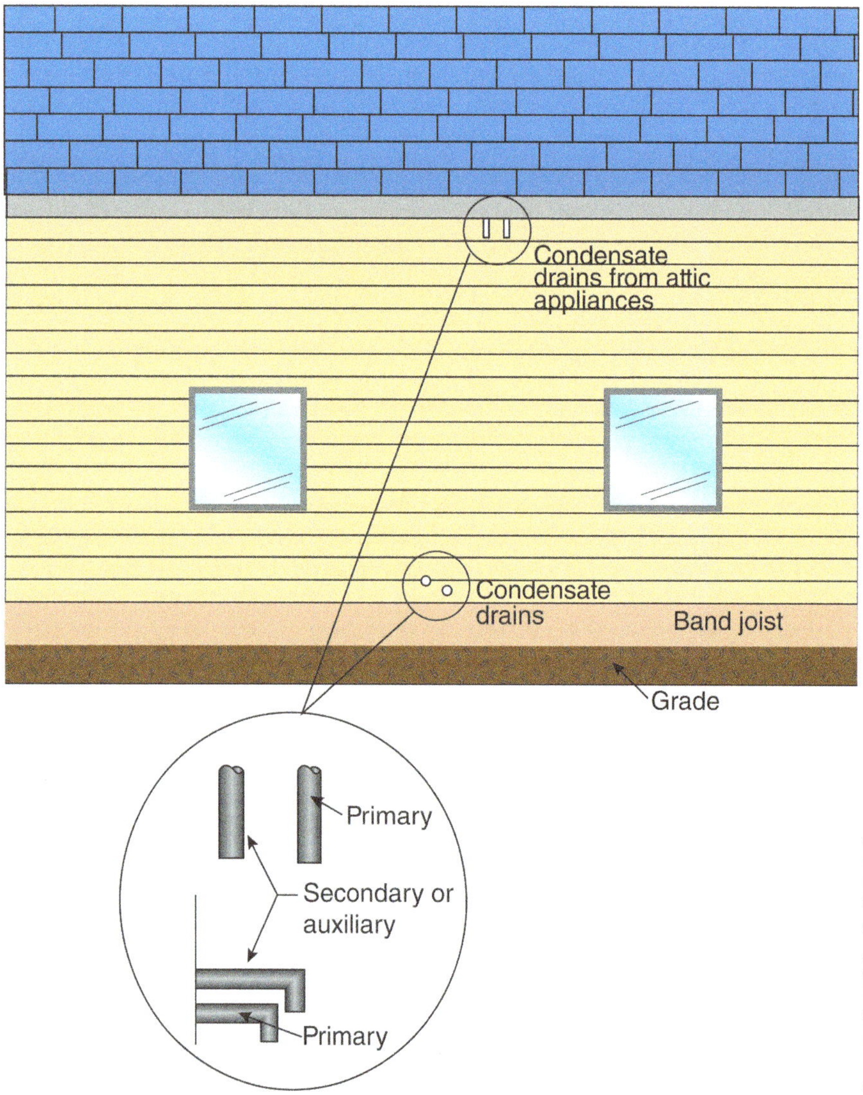

Labeling of concealed condensate piping termination

CHANGE SIGNIFICANCE: Section 307.2 requires a primary condensate drainage system, and Section 307.5 requires an auxiliary drain pan and drain where overflow of the primary drain would cause damage to the building. Where the two drain lines serving the same remote appliance are not readily observable, one would not know which drain line was which. For example, if the two drain lines came from an attic space and terminated through an outside wall, one could not discern which was the primary drain. If condensate is flowing from the auxiliary drain, that is an indication of a failed primary drain line and immediate attention would be needed to correct the problem. If the two drain lines were not labeled as to their purpose, it would not be possible to distinguish between normal primary drain line flow and abnormal auxiliary drain line flow. The required marking is intended to be permanent; therefore, it must be suitable for the location so as to remain visible for the life of the system. Duct tape and a felt tip pen marking would not survive long exposed to the weather.

402.7, 202

Press-Connect Joint

CHANGE TYPE: Modification

CHANGE SUMMARY: The code now recognizes press-connect joints as suitable for high pressure (over 5 psig) applications indoors.

2021 CODE: **402.7 Maximum operating pressure.** The maximum operating pressure for piping systems located inside buildings shall not exceed 5 pounds per square inch gauge (psig) (34 kPa gauge) except where one or more of the following conditions are met:

1. The piping joints are welded or brazed.
2. The piping is joined by fittings listed to ANSI LC-4/CSA 6.32 and installed in accordance with the manufacturer's instructions.

(Items 3 through 8 are unchanged)

SECTION 202
GENERAL DEFINITIONS

PRESS-CONNECT JOINT. A permanent mechanical joint incorporating an elastomeric seal or an elastomeric seal and corrosion-resistant grip or bite ring. The joint is made with a pressing tool and jaw or ring approved by the fitting manufacturer.

CHANGE SIGNIFICANCE: A new type of piping system joint is now allowed for gas piping systems inside buildings that operate at pressures greater than 5 psig. As described in the definition of "press-connect joint" and as can be seen in the photos, the fittings for steel pipe have a "bite ring" with teeth that indent in the steel pipe wall so the fitting is impossible to remove or displace. Press-connect fittings for gas piping systems were already allowed under Sections 403.10.1 and 403.10.2, but were not allowed if the gas pressure in the system exceeded 5 psig. Press-connect fittings form a seal by means of a compressed O-ring that is specially formulated to tolerate high temperatures and extreme conditions dictated by the product standard ANSI LC-4/CSA 6.32. Press-connect joints are common for copper plumbing piping and are gaining popularity for steel piping.

Photo courtesy of Viega LLC

Press-connect fitting with bite ring used with Sch 40 and Sch 10 steel pipe

403.8.3

Threaded Joint Sealing

CHANGE TYPE: Modification

CHANGE SUMMARY: The text was revised to require the use of thread joint sealants (aka, joint compounds, pipe dope, pipe tape).

2021 CODE: **403.8.3 Threaded joint ~~compounds~~ sealing.** Threaded joints shall be made using a thread joint sealing material. Thread joint sealing materials ~~compounds~~ shall be nonhardening and shall be resistant to the ~~action of liquefied petroleum gas or to any other~~ chemical constituents of the gases to be conducted through the piping. Thread joint sealing materials shall be compatible with the pipe and fitting materials on which the sealing materials are used.

CHANGE SIGNIFICANCE: The code addressed pipe thread sealants but never required them to be used. Thread sealants act primarily as a lubricant to allow the threads to make up tight to form a metal-to-metal seal, and any imperfections or voids in the threads are filled in by the thread sealant material. The most common thread sealants used today are pastes made with PTFE (Teflon) and Teflon tapes. Sealing compounds must not be hardening so that pipe joints can be disassembled when necessary, shrinkage of the sealant is minimized and the material can be resilient to vibration and movement.

LP-gas can react with some pipe sealants, and this might also be the case for other gases; therefore, the text simply requires that the sealant be compatible with whatever is conveyed in the piping. The thread sealant used must be suitable for the pipe and fitting materials used. Some thread sealants can attack some pipe and fitting materials. This is more of an issue with plumbing piping, which often has plastic threads, while gas piping always involves metallic threads. Thread sealant container labels will state what pipe and fitting materials are suitable or not suitable for use with the sealant.

Photo courtesy of gettyimages.com/Vudhikul Ocharoen

Threaded joint pipe tape

404.5

Fittings in Concealed Locations

CHANGE TYPE: Modification

CHANGE SUMMARY: This change addresses an omission of threaded fittings that were intended to be allowed in concealed locations, yet not specifically mentioned in the text.

2021 CODE: **404.5 Fittings in concealed locations.** Fittings installed in concealed locations shall be limited to the following types:

1. Threaded elbows, tees, couplings, plugs and caps.
2. Brazed fittings.
3. Welded fittings.
4. Fittings listed to ANSI LC-1/CSA 6.26 or ANSI LC-4/CSA 6.32.

CHANGE SIGNIFICANCE: Threaded (screwed) fittings, such as those used with steel piping, are allowed to be concealed, except for unions. Item 1 of this section lists the threaded fittings that are allowed to be concealed. Unions are not in the list and are, therefore, prohibited in concealed locations. There was no reason to exclude threaded plugs and threaded caps, so that omission was corrected. Left/right couplings can be viewed as a type of union; however, they are nonetheless threaded couplings and are not prohibited.

International Code Council

Threaded fittings permitted in concealed spaces

503.5.6.1 Chimney Lining

CHANGE TYPE: Modification

CHANGE SUMMARY: The exception for existing chimneys, previously intended to avoid a hardship, is considered to be outdated and has been deleted.

2021 CODE: **503.5.6.1 Chimney lining.** Chimneys shall be lined in accordance with NFPA 211.

> ~~**Exception:** Where an existing chimney complies with Sections 503.5.6 through 503.5.6.3 and its sizing is in accordance with Section 503.5.5, its continued use shall be allowed where the appliance vented by such chimney is replaced by an appliance of similar type, input rating and efficiency.~~

CHANGE SIGNIFICANCE: The exception allowed an existing unlined chimney to vent replacement appliances, where such replacement appliances were similar in type, Btu/h input rating and thermal efficiency. A replacement appliance installation is treated the same as the initial appliance installation, meaning that it is new work that must comply with the code that is currently in effect for new work. The exception applied to all appliances, but for this discussion, let's consider water heaters. For

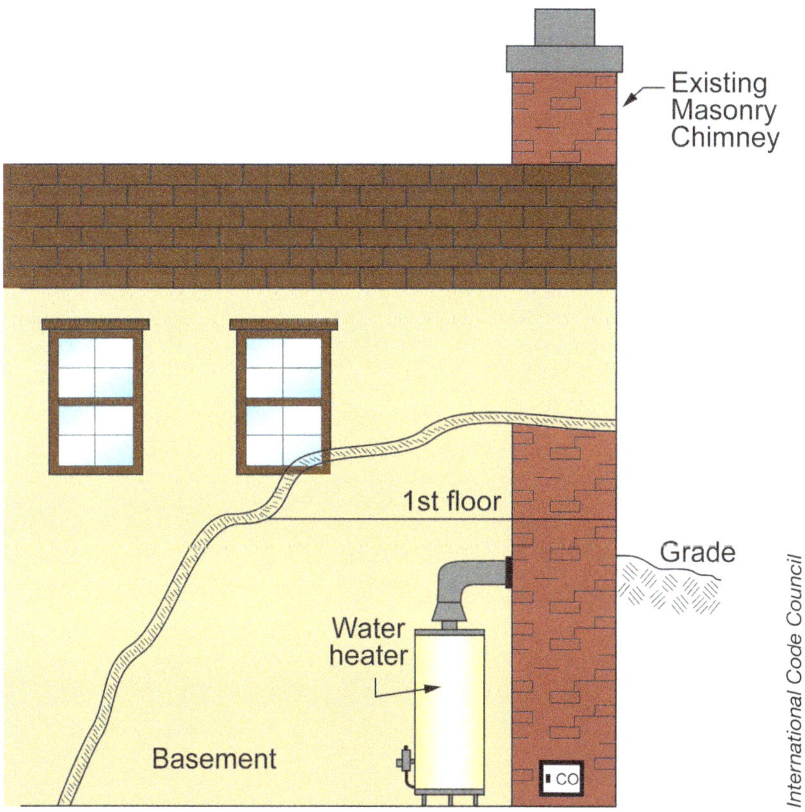

Code-compliant chimney lining required for replacement appliance

example, a replacement water heater installation is required to be served by a code-compliant venting system, just as the initial water heater was. However, this exception allowed a non-code-compliant chimney to serve a new water heater installation, in conflict with a basic principle of the code, the appliance installation instructions and the appliance listing. The reason for this was that many existing buildings, typically homes, had old unlined existing chimneys and it would be a financial hardship to require the chimney to be lined if the water heater had to be replaced, especially if the previous water heater was apparently being vented properly. The exception was based on the assumption that the previous water heater was venting properly. If it was not, the replacement appliance would not be venting properly either. Because newer appliances change design, input ratings and efficiency, it is unlikely that the replacement appliance will be "similar" to the previous appliance. Serviceable unlined chimneys have become quite rare in an era where high-efficiency appliances dominate. Also, consider that the exception would rarely if ever be applicable because it required the existing unlined chimney to be correctly sized and in good condition, which would be exceedingly rare. For all of the reasons above, the time had come for the hardship exception to be deleted.

503.8

Venting System Terminal Clearances

CHANGE TYPE: Modification

CHANGE SUMMARY: Though-the-wall vent terminal clearance distances have been placed in a new table with a corresponding figure for ease of use.

2021 CODE: **503.8 Venting system ~~termination location~~ terminal clearances.** The clearances for through-the-wall direct-vent and non-direct-vent terminals shall be in accordance with Table 503.8 and Figure 503.8.

> **Exception:** The clearances in Table 503.8 shall not apply to the combustion air intake of a direct-vent appliance

~~The location of venting system terminations shall comply with the following (see Appendix C):~~

1. ~~A mechanical draft venting system shall terminate not less than 3 feet (914 mm) above any forced-air inlet located within 10 feet (3048 mm).~~

   ~~**Exceptions:**~~

   1. ~~This 83provision shall not apply to the combustion air intake of a direct-vent appliance.~~
   2. ~~This provision shall not apply to the separation of the integral outdoor air inlet and flue gas discharge of listed outdoor appliances.~~

2. ~~A mechanical draft venting system, excluding direct-vent appliances, shall terminate not less than 4 feet (1219 mm) below, 4 feet (1219 mm) horizontally from, or 1 foot (305 mm) above any door, operable window or gravity air inlet into any building. The bottom of the vent terminal shall be located not less than 12 inches (305 mm) above finished ground level.~~
3. ~~The clearances for through-the-wall, direct-vent terminals shall be in accordance with Table 503.8. The bottom of the vent terminal and the air intake shall be located not less than 12 inches (305 mm) above finished ground level.~~
4. ~~Through-the-wall vents for Category II and IV appliances and noncategorized condensing appliances shall not terminate over public walkways or over an area where condensate or vapor could create a nuisance or hazard or could be detrimental to the operation of regulators, relief valves or other equipment. Where local experience indicates that condensate is a problem with Category I and III appliances, this provision shall also apply. Drains for condensate shall be installed in accordance with the appliance and vent manufacturers' instructions.~~

~~5. Vent systems for Category IV appliances that terminate through an outside wall of a building and discharge flue gases perpendicular to the adjacent wall shall be located not less than 10 feet (3048 mm) horizontally from an operable opening in an adjacent building. This requirement shall not apply to vent terminals that are 2 feet (607 mm) or more above or 25 feet (7620 mm) or more below operable openings.~~

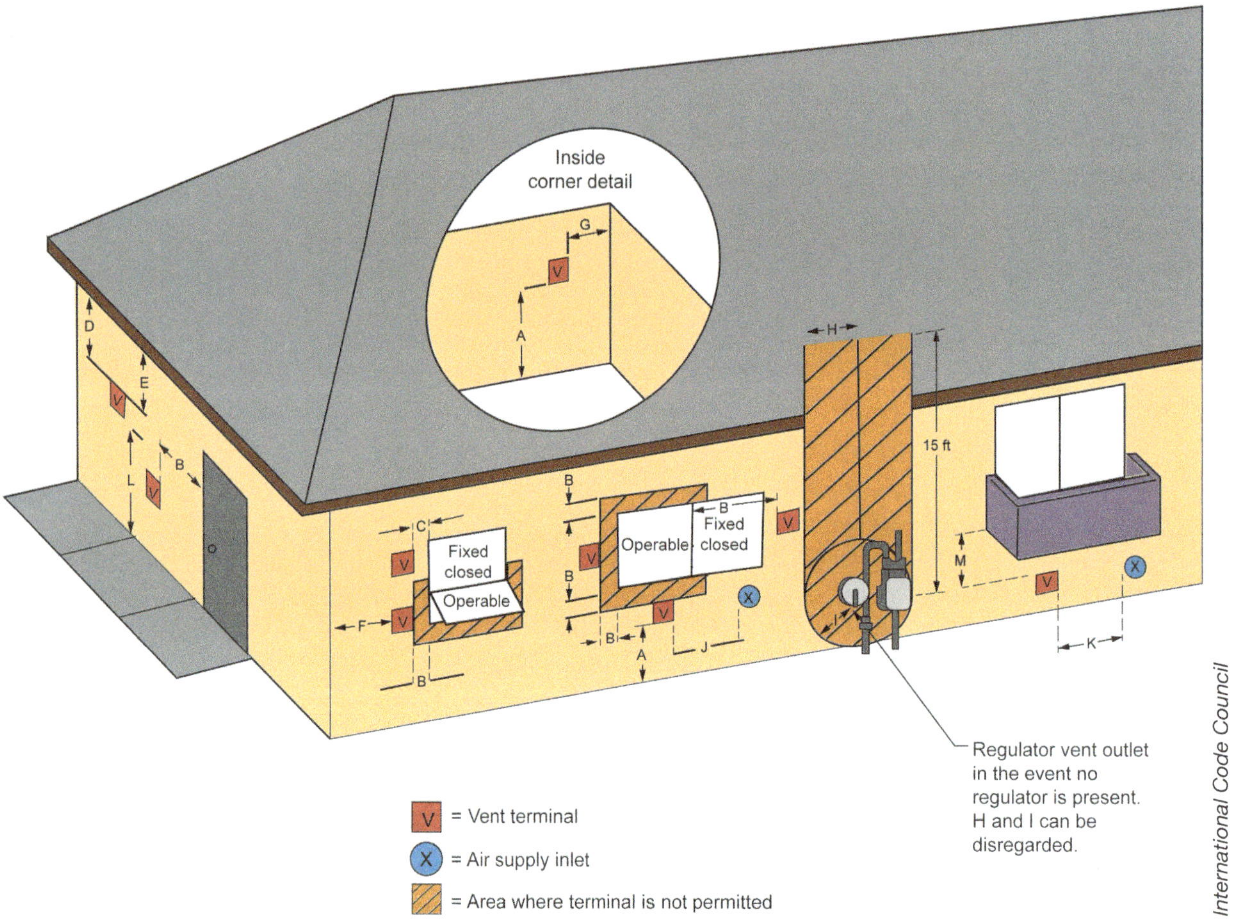

Figure 503.8 Through the wall vent terminal clearances

~~**TABLE 503.8 Through-The-Wall Direct-Vent Termination Clearances**~~

~~Direct-Vent Appliance Input Rating (Btu/hr)~~	~~Through-The-Wall Vent Terminal Clearance From Any Air Opening Into The Building (Inches)~~
~~< 10,000~~	~~6~~
~~≥ 10,000 ≤ 50,000~~	~~9~~
~~> 50,000 ≤ 150,000~~	~~12~~
~~> 150,000~~	~~In accordance with the appliance manufacturer's instructions and not less than the clearances specified in Section 503.8, Item 2~~

~~For SI: 1 inch = 25.4 mm, 1 Btu/h = 0.2931 W.~~

TABLE 503.8 Through-the-Wall Vent Terminal Clearances

Figure Clearance	Clearance Location	Minimum Clearances for Direct-Vent Terminals	Minimum Clearances for Nondirect-vent Terminals
A	Clearance above finished grade level, veranda, porch, deck, or balcony	12 in.	
B	Clearance to window or door that is openable	6 in.: Appliances ≤ 10,000 Btu/hr 9 in.: Appliances > 10,000 Btu/hr ≤ 50,000 Btu/hr 12 in.: Appliances > 50,000 Btu/hr ≤ 150,000 Btu/hr Appliances > 150,000 Btu/hr, in accordance with the appliance manufacturer's instructions and not less than the clearances specified for nondirect-vent terminals in Row B	4 ft. below or to side of opening or 1 foot above opening
C	Clearance to nonopenable window	None unless otherwise specified by the appliance manufacturer	
D	Vertical clearance to ventilated soffit located above the terminal within a horizontal distance of 2 feet (61 cm) from the center line of the terminal	None unless otherwise specified by the appliance manufacturer	
E	Clearance to unventilated soffit	None unless otherwise specified by the appliance manufacturer	
F	Clearance to outside corner of building	None unless otherwise specified by the appliance manufacturer	
G	Clearance to inside corner of building	None unless otherwise specified by the appliance manufacturer	
H	Clearance to each side of center line extended above regulator vent outlet	3 ft. up to a height of 15 ft. above the regulator vent outlet	
I	Clearance to service regulator vent outlet in all directions	3 ft. for gas pressures up to 2 psi; 10 ft. for gas pressures above 2 psi	
J	Clearance to nonmechanical air supply inlet to building and the combustion air inlet to any other appliance	Same clearance as specified for Row B	
K	Clearance to a mechanical air supply inlet	10 ft. horizontally from inlet or 3 ft. above inlet	
L	Clearance above paved sidewalk or paved driveway located on public property	7 ft. and shall not be located above public walkways or other areas where condensate or vapor can cause a nuisance or hazard	
M	Clearance to underside of veranda, porch, deck, or balcony	12 in. where the area beneath the veranda, porch, deck or balcony is open on not less than two sides. The vent terminal is prohibited in this location where only one side is open.	

For SI units, 1 in. = 25.4 mm, 1 ft. = 304.8 mm, 1 Btu/hr = 0.293 W

CHANGE SIGNIFICANCE: The reformatting of Section 503.8 is not meant to make any technical changes to the though-the-wall vent terminal clearance provisions. Reorganizing the content and placing all of the clearance requirements into a single table makes the information easier to locate and understand than the previous layout of the material. In addition, Figure 503.8 illustrates the location and clearances for the vent terminations. The identifying letter in the table corresponds to the lettered locations in the figure. The figure is similar to the figure that has appeared in Appendix C. The new figure captures all of the terminal locations with the corresponding identifying letters from the table.

503.10.7

Vent Connector Junctions

CHANGE TYPE: Addition

CHANGE SUMMARY: New text addresses the juncture of appliance vent connectors and the required method.

2021 CODE: **503.10.7 Connector junctions.** Where vent connectors are joined together, the connection shall be made with a tee or wye fitting.

CHANGE SIGNIFICANCE: This new requirement was added to address, primarily, single-wall vent connectors that are joined together before connecting to a chimney, B-vent or Type L vent. An age-old practice was to cut a hole in the side or bottom of a single-wall metal pipe vent connector to join a smaller vent connector pipe to it. For example, a 6-inch vertical single-wall pipe vent connector serving a boiler or furnace has a hole cut in the side of it and a 3-inch single-wall pipe vent connector serving a water heater is inserted into the hole and fastened in place by bent tabs that are screwed or riveted to the larger vent connector pipe.

Such connection is not substantial or secure, allows considerable leakage that can affect draft, creates an obstruction inside the larger vent connector, typically ignores connector sizing criteria and is, at best, poor practice. This new section requires the use of manufactured tee or wye fittings. This has not been an issue with double-wall connectors because they are necessarily joined with tee and wye fittings designed for the purpose.

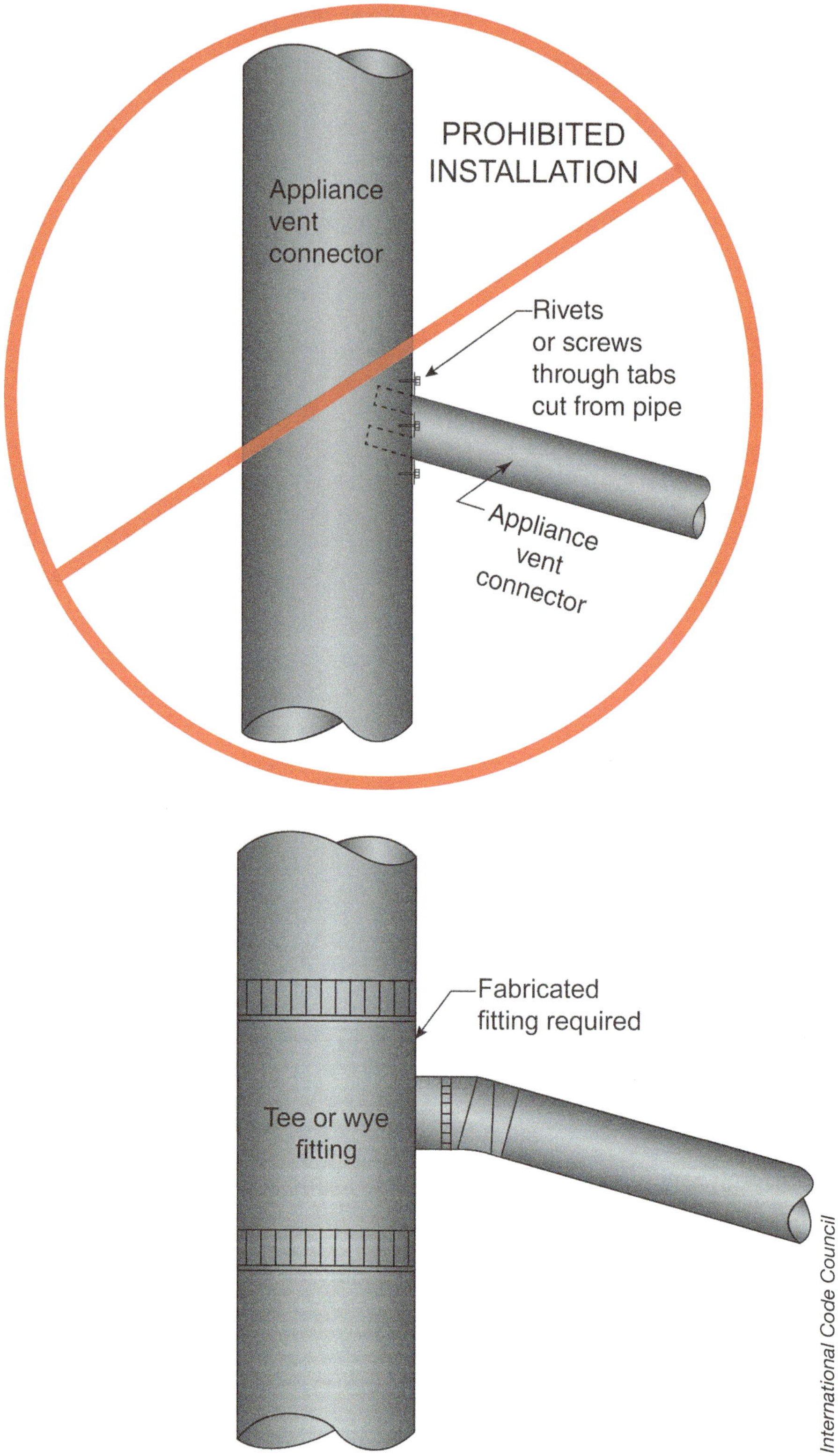

Vent connectors joined by tee or wye fittings

614.7

Clothes Dryer Makeup Air

CHANGE TYPE: Clarification

CHANGE SUMMARY: This revision separated two unrelated requirements for clothes dryer makeup air into separate sections to clarify the applicability of each.

2021 CODE: **614.7 Makeup air.** Installations exhausting more than 200 cfm (0.09 m^3/s) shall be provided with makeup air. ~~Where a closet is designed for the installation of a clothes dryer, an opening having an area of not less than 100 square inches (645 mm^2) for makeup air shall be provided in the closet enclosure, or makeup air shall be provided by other approved means.~~

614.7.1 Closet Installation. Where a closet is designed for the installation of a clothes dryer, an opening having an area of not less than 100 square inches (645 mm^2) for makeup air shall be provided in the closet enclosure, or makeup air shall be provided by other approved means.

CHANGE SIGNIFICANCE: Section 614.7 requires all dryers exhausting greater than 200 cfm to be provided with a makeup air supply because of the concern for creating negative pressure in the space and starving the dryer for airflow. This requirement is distinct from the requirement for providing an opening in closet enclosures that contain dryers. Section 614.7 now addresses only the requirement to provide makeup air for dryers that exceed the 200 cfm threshold, and Section 614.7.1 addresses the requirement to have an opening in the closet enclosure to allow the makeup air into the closet. Without the opening into the closet, the dryer would be isolated from its source of makeup air and the closet would experience significant negative pressure. The requirement of Section 614.7.1 must necessarily apply to all dryers, regardless of the cfm exhaust rate, while Section 614.7 of the 2018 code appeared to require the closet opening only for dryers that exhaust over 200 cfm.

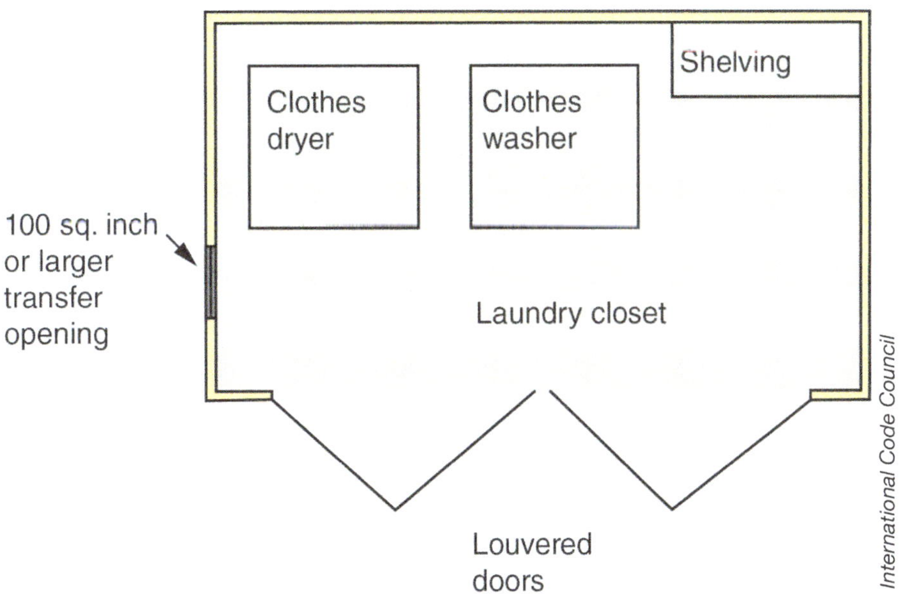

Makeup air for a closet installation of a dryer

618.6

Return Air from Mechanical Room

CHANGE TYPE: Modification

CHANGE SUMMARY: New text clarifies the intent to prohibit pulling return air from the mechanical room.

2021 CODE: **618.6 Furnace plenums and air ducts.** Where a furnace is installed so that supply ducts carry air circulated by the furnace to areas outside of the space containing the furnace, the return air shall be handled by a duct(s) sealed to the furnace casing and terminating outside of the space containing the furnace. <u>Return air shall not be taken from the mechanical room containing the furnace.</u>

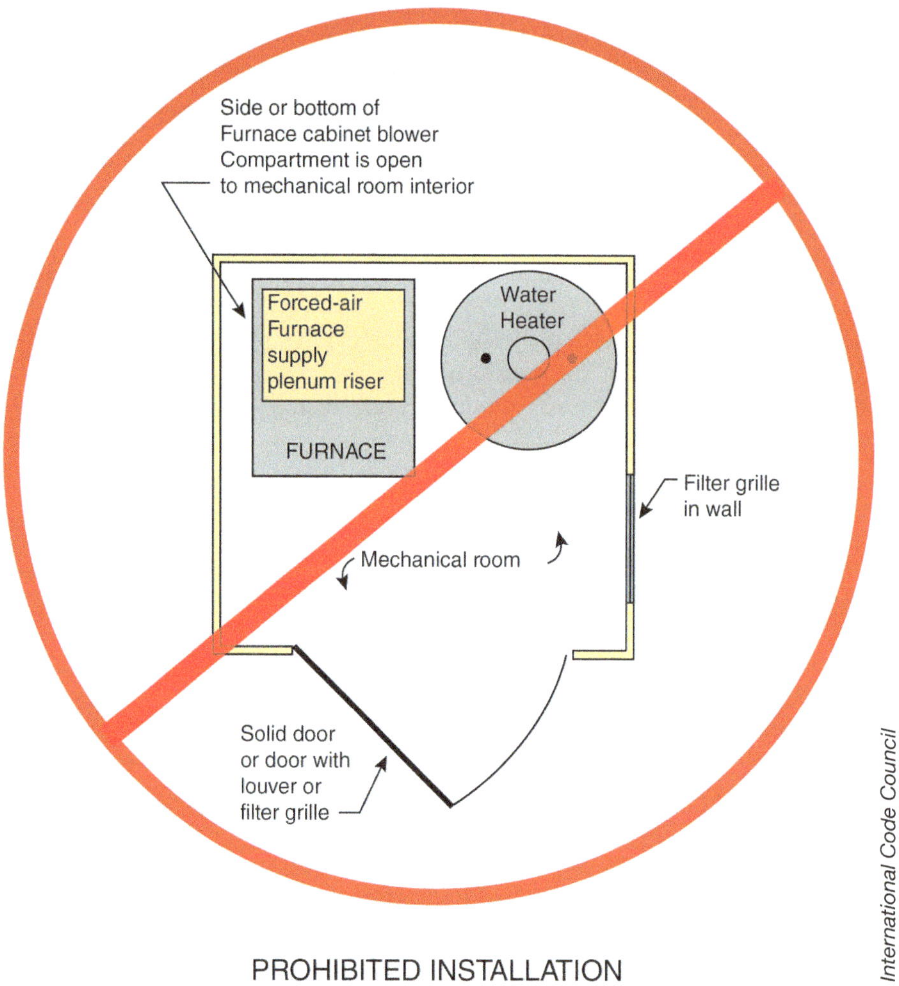

PROHIBITED INSTALLATION

Prohibited installation of return air taken from the mechanical room containing the furnace

CHANGE SIGNIFICANCE: The wording of this section was not sufficiently clear to all readers and furnace installations have often been in violation of this section. This section intends to prevent a significant hazard. The added text explicitly states the intent that the furnace must not draw air from the closet or mechanical room that contains the furnace. Many dwelling units have been constructed with furnaces installed in closets, commonly with a water heater, and the furnace return air inlet opening is open to the closet interior. Return air from the dwelling is drawn through a louvered door or grille in the closet enclosure and passes through the closet interior and into the furnace return air opening. Of course, this creates a strong negative pressure within the closet space as the return airflow overcomes the resistance of the louvered door or wall grille and any filters in such. Negative pressures in the closet space can interfere with the furnace and water heater vents, causing flue gas spillage, appliance malfunction and entrainment of flue gases in the return airflow. This section addresses furnaces; however, the same hazard can exist if an electric furnace or heat pump or cooling system air handler is installed in a closet with a gas-fired water heater or boiler.

The added text makes it clear that return air for the furnace must be ducted to the furnace from outside the furnace closet to prevent the creation of negative pressure within the closet. It is important to note that this section grants no relief or exception for direct-vent or power-vented appliances. Direct-vent and power-vented appliances are less affected by negative pressures than open combustion chamber natural draft appliances, but, direct-vent and power-vented appliances are not immune to negative pressures.

623.2

Commercial Cooking Appliances

CHANGE TYPE: Modification

CHANGE SUMMARY: An unnecessary exception for commercial cooking appliances within dwelling units has been deleted.

2021 CODE: 623.2 Prohibited location. Cooking appliances designed, tested, listed and labeled for use in commercial occupancies shall not be installed within dwelling units or within any area where domestic cooking operations occur.

> **~~Exceptions~~ Exception:**
>
> ~~1.~~ Appliances that are also listed as domestic cooking appliances.
>
> ~~2. Where the installation is designed by a licensed Professional Engineer, in compliance with the manufacturer's installation instructions.~~

CHANGE SIGNIFICANCE: The exception was created because some consumers were demanding commercial cooking appliances in their homes as this has become a popular trend. Also, at least one state government modified the adopted IFGC to permit commercial cooking appliances to be installed in dwellings if an engineer designed the job. The marketplace has taken care of this demand, because appliance manufacturers currently offer many commercial-style appliances that are duel listed as both commercial and household appliances. This dual listing is already addressed in Exception 1. There is no reason to risk occupant safety by installing commercial-only appliances in dwellings when there are many appliances listed for dwelling (household) installation and have the commercial features and capacities that consumers want.

Exception 2 permitted cooking appliances listed only for commercial applications to be installed in a dwelling if a licensed professional engineer designed the kitchen installation and followed the manufacturer's installation instructions. This is a contradictory statement, because the manufacturer's instructions generally prohibit the installation of commercial appliances in dwellings; thus, the engineer could not design an installation that complied with the appliance manufacturer's instructions. This is also contrary to Section 301.3, which requires appliances to be listed and labeled for the application in which they are used. Commercial

International Code Council

Commercial-type domestic range

ranges that are not listed for domestic (household) use are not listed and labeled for use in a dwelling; therefore, it is a code violation whether or not it has the blessing of the design engineer. The typical appliance nameplates state that a commercial appliance is not intended for domestic use, so installation in a dwelling would be a violation of the manufacturer's instructions as well as Section 301.3. Commercial cooking appliances are not suitable or even safe within dwellings for multiple reasons, including, for example, that they require large clearances to combustibles, they might have much higher input ratings and surface temperatures, they might require non-combustible floors, they might lack some safety features required for domestic appliances and they might require exhaust systems similar to what is required in restaurants. This conflicted exception is not necessary because of the allowance of Exception 1 and the allowance in Chapter 1 for the code official to grant alternative approval for installations that meet the spirit and intent of the code and that are deemed equivalent to what the code specifies. For example, there have likely been cases where the code official has approved the installation of commercial-only appliances in extraordinary homes where the kitchen was built and operated like a restaurant and children were not present.